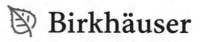

Frontiers in Mathematics

Advisory Board

Leonid Bunimovich (Georgia Institute of Technology, Atlanta)
William Y. C. Chen (Nankai University, Tianjin)
Benoît Perthame (Sorbonne Université, Paris)
Laurent Saloff-Coste (Cornell University, Ithaca)
Igor Shparlinski (The University of New South Wales, Sydney)
Wolfgang Sprößig (TU Bergakademie Freiberg)
Cédric Villani (Institut Henri Poincaré, Paris)

More information about this series at http://www.springer.com/series/5388

Vladislav V. Kravchenko

Direct and Inverse Sturm-Liouville Problems

A Method of Solution

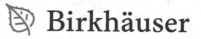

Vladislav V. Kravchenko
Department of Mathematics
Cinvestav
Querétaro, Mexico

ISSN 1660-8046   ISSN 1660-8054   (electronic)
Frontiers in Mathematics
ISBN 978-3-030-47848-3   ISBN 978-3-030-47849-0   (eBook)
https://doi.org/10.1007/978-3-030-47849-0

Mathematics Subject Classification: 34-02, 34A25, 34A45, 34A55, 34B05, 34B08, 34B09, 34B24, 34B40, 34L05, 34L10, 34L16, 34L25, 34L40, 35Q53, 47A40, 47E05, 65L05, 65L09, 65L15, 81U40

This book is published under the imprint Birkhäuser, www.birkhauser-science.com, by the registered company Springer Nature Switzerland AG.
The registered company address is: Gewerbestrasse 11, 6330 Cham, Switzerland

To Kira

Preface

This book presents recent results on applications of the transmutation (transformation) operator method to practical solution of direct and inverse Sturm–Liouville problems on finite and infinite intervals. It is intended for use by mathematicians interested in the theory and numerics of direct and inverse Sturm–Liouville problems, and the efficient construction of the transmutation operators, as well as by physicists and engineers interested in simple and practical methods for fast and accurate solution of such problems.

The author did not set for himself the task of making this short book self-contained. A number of theorems are presented without proof, but always with appropriate references. The criterion was to include the complete proofs of those results that have not yet appeared in books.

Thus, the main material of the book is the exposition of the approach developed by the author and colleagues during the last several years to a practical solution of the computationally challenging direct and inverse Sturm–Liouville problems on finite and infinite intervals. For this purpose, many theoretical results were used and developed, and the methods described below offer much more than the numerical solution of difficult problems. In all cases, we derive analytical representations of solutions and of integral kernels of the transmutation operators involved, which leads to the possibility of working efficiently with the obtained solutions. Different phenomena, such as dispersion or asymptotic behaviour, are studied easier when the solution of the problem is available in an analytical form.

The methods and results presented in this book can be used in models coming from a wide spectrum of applied sciences.

The author expresses his gratitude to Conacyt (Mexico) for partial support of this work via the project 284470 and acknowledges the support by the Regional Mathematical Center of the Southern Federal University, Rostov-on-Don, Russia.

Querétaro, Mexico   Vladislav V. Kravchenko

Contents

Introduction

1

Since the pioneering work of D. Bernoulli, J. d'Alembert, L. Euler, J. Fourier and later on of S. D. Poisson, Ch. Sturm and J. Liouville [131], the theory of Sturm-Liouville problems is an integral part of the professional preparation of mathematicians, physicists and engineers, and at the same time an important and actively developing research field.

With the work of V. A. Ambartsumyan [13] and G. Borg [33], the theory of inverse Sturm-Liouville problems began to develop rapidly and took an important place in general spectral theory, with numerous applications in mathematics, mechanics, physics, engineering and other applied sciences. Various aspects of direct and inverse Sturm-Liouville problems are covered in fundamental books, such as [31, 51, 69, 82, 127, 128, 133, 139, 150, 162, 163, 172, 174], and many others. This is a beautiful piece of modern mathematics, important for development of real-world sciences.

The theory of direct and inverse Sturm-Liouville problems is quite well developed, although the lack of practical methods for their study and solution is evident even from standard textbooks on differential equations and methods of mathematical physics. For example, while the basic properties of the eigenvalues and eigenfunctions of a regular Sturm-Liouville problem are formulated and proved for a general Sturm-Liouville equation with variable coefficients, the practical solution of Sturm-Liouville problems is studied on equations with constant coefficients only, and no reasonably practical method is taught for solving Sturm-Liouville equations with variable coefficients except probably for such purely numerical techniques as the finite difference method.

The aim of this short book is to present such practical methods for solving the Sturm-Liouville equations and direct and inverse Sturm-Liouville problems. As the reader will see leafing through the book, behind all the new constructions and results presented here stands the notion of the transmutation (or transformation) operator. In relation with linear differential operators, it appeared first in the work by J. Delsarte [60, 61] and later on was

V. V. Kravchenko, *Direct and Inverse Sturm-Liouville Problems*, Frontiers in Mathematics, https://doi.org/10.1007/978-3-030-47849-0_1

developed in numerous publications. We refer to the books [28, 42–44, 65, 86, 112, 127, 133, 159, 160, 164] and to the useful survey [158], which offers an impressive bibliography on the subject.

Roughly speaking, the transmutation operators studied in this book relate solutions of Sturm-Liouville equations with variable coefficients to solutions of the elementary equation $y'' + \lambda y = 0$, where λ is constant. The knowledge of an appropriate transmutation operator leads to a complete solution of the Sturm-Liouville equation. Since the work [140] it is even known that such transmutation operators are realizable in the form of Volterra integral operators of the second kind with kernels independent of the spectral parameter of the Sturm-Liouville equation. The mere fact of the existence of the transmutation operators in such a convenient form allowed the mathematicians to advance the theory of direct and inverse Sturm-Liouville problems. Perhaps, V. A. Marchenko was the first to apply systematically transmutation operators as an important mathematical tool in the theory of direct and inverse spectral problems [132, 133].

Moreover, in the important work [74] I. M. Gelfand and B. M. Levitan discovered that as a key to the solution of the inverse Sturm-Liouville problem one can take the integral kernel of the transmutation operator. They showed that the kernel satisfies an integral equation which can be constructed from the given spectral data of the inverse problem. The equation is known as the Gelfand-Levitan equation and apparently [74] solves the inverse Sturm-Liouville problem by reducing it to a Fredholm integral equation of second kind. However, that was not the end of the story of inverse Sturm-Liouville problems, in particular because the attempts to use the Gelfand-Levitan equation directly for numerical solution of inverse Sturm-Liouville problems have not been fruitful (a discussion of the difficulties arising can be found in [130]).

Since the magic of the transmutation operators is required in a wide variety of fields, attempts were made to approximate their integral kernels. The natural approach via the method of successive approximations was employed, e.g., in [28] and [56]. In [34] an attempt of constructing the Fourier series of the kernels, leading to a system of equations for the Fourier coefficients, was reported. In [115, 116] the transmutation integral kernels were approximated by so-called transmuted wave polynomials (see [89] and [92]). A more complete and satisfactory result was obtained in [106], where an exact Fourier-Legendre functional series representation for the transmutation integral kernels was obtained with a simple recurrent integration procedure for the coefficients of the series. The substitution of the series into the representation of solutions of the Sturm-Liouville equation led to a new series representation for solutions in the form of so-called Neumann series of Bessel functions (NSBF). Moreover, since the NSBF representations are obtained via the transmutation operator, they enjoy a very singular uniformity feature. Namely, the remainder of the series admits estimates independent of the real part of the square root of the spectral parameter. That is, if $\rho := \sqrt{\lambda}$ and λ is the spectral parameter in the Sturm-Liouville equation, the estimate of the remainder of the NSBF does not depend on $\operatorname{Re} \rho$. For example, suppose that we need to solve a regular Sturm-Liouville problem for the equation

$$- y'' + q(x)y = \lambda y. \tag{1.1}$$

Then a finite number of negative eigenvalues may exist, while there always exists an infinite sequence of positive eigenvalues going to infinity. The estimate mentioned above allows one to compute very large sets of the eigendata (eigenvalues and eigenfunctions) with a uniform accuracy and in no time.

That work on the exact functional series representations for integral transmutation kernels received several continuations. In [118] similar results were obtained for the general Sturm-Liouville equation while in [106] this was done for (1.1). In [119] similar results were obtained for the perturbed Bessel equation (see also [111]). In [49, 91] and [99] the results of [106] were applied to other interesting problems by exploiting the observation that in other terms the above NSBF representations offer the way to calculate the images of the functions $e^{\pm in\rho x}$, $n = 0, 1, \ldots$ under the action of the transmutation operator. The possibility to derive other functional series representations for the integral transmutation kernels, leading to different interesting series representations for solutions of (1.1), was explored in [120] and [102]. All these continuations of the work [106] are out of the scope of this book.

While the direct Sturm-Liouville problems on finite intervals can be solved efficiently with the use of the NSBF representations, there remained the question of how to apply the Fourier-Legendre series representation to the solution of the inverse Sturm Liouville problems. A way to do it was proposed in [103], where for this purpose the Gelfand-Levitan equation was used. The substitution of the Fourier-Legendre series for the transmutation kernel into the Gelfand-Levitan equation eventually led to a linear system of algebraic equations for the coefficients of the series, and a crucial observation was that for a complete solution of the inverse problem it is enough to calculate only the very first coefficient. This is sufficient for recovering the potential $q(x)$ in (1.1) and the constants in the Sturm-Liouville boundary conditions. Unlike existing numerical methods for solving the inverse Sturm-Liouville problem on a finite interval, the method developed in [103] is not iterative. It reduces the problem directly to a linear system of algebraic equations from which, in fact, only the very first entry of the solution vector needs to be found. This means that solving a truncated system with very few equations leads to a satisfactory solution of the inverse problem.

The same approach worked successfully when applied to the inverse Sturm-Liouville problem on the half-line [58] and to the inverse quantum scattering problem on the half-line [85]. These two problems can be reduced to corresponding Gelfand-Levitan equations for the integral kernels of transmutation operators with boundary conditions at the origin, but with different input kernels containing the spectral (or the scattering) data of the problem. However, the inverse scattering problem on the line required some additional ideas, because for its solution the integral kernel $A(x, t)$ of the transmutation operator with a condition at infinity is required. An approach similar to the one described above was proposed in [104], where the main idea came from the observation that the first coefficient of a functional series representation of the transmutation kernel is sufficient for recovering the potential. It turned out quite easy to observe that the kernel $A(x, t)$ admits a certain Fourier-Laguerre series representation; although it was not clear how to compute its coefficients, it ended up being straightforward how the potential $q(x)$ can be recovered from the first coefficient. Thus, we managed to develop essentially the same approach, and construct a direct and simple method for solving the inverse scattering problem on the line.

In a follow-up publication [59] a recurrent integration procedure for calculating the coefficients of the Fourier-Laguerre series of the kernel $A(x, t)$ was derived, leading to a new series representation for the so-called Jost solution. This representation allowed us to reduce, e.g., the direct Sturm-Liouville problem on the half-line, to a computation of quite simple expressions (which are essentially power series) along the unit circle and one of its diameters. The numerical solution of the Sturm-Liouville problem on the half-line for a short-range potential is quite a challenge. It is known to be difficult and computationally expensive. We refer the reader to the book [142], where in the last section the author explains the difficulty of the problem, and to more recent papers [71] and [72]. We mention the work [11], where a method for computing the Jost function was proposed. The representation of the Jost solution obtained in [59] and presented in Chap. 10, in fact, trivializes this problem, allowing one in a fraction of a second on a laptop to plot the derivative of the spectral density function with remarkable accuracy and on an arbitrarily large interval in λ. Moreover, writing a corresponding computer program is an easy task that can be assigned as a homework exercise to an undergraduate student.

In this book the approach based on the transmutation operators and their efficient construction is presented. It allows one to obtain analytical representations of solutions of Sturm-Liouville equations and solve in practice direct and inverse Sturm-Liouville problems on finite and infinite intervals. The book consists of four parts. In the first one some necessary definitions and results are given, as well as the formulation of several typical direct and inverse Sturm-Liouville problems. In the second part functional series representations for the integral transmutation kernels and for the solutions of the Sturm-Liouville equation are derived. Here, in Chap. 8 some necessary facts on transmutation operators are collected. In Chap. 9 the Fourier-Legendre series for the transmutation operators and the corresponding NSBF for solutions are derived. In Chap. 10 the Fourier-Laguerre series for the transmutation kernel $A(x, t)$ and a functional series representation for the Jost solution are presented. The third part is devoted to methods of solution of direct Sturm-Liouville problems, first on finite and then on infinite intervals. In the fourth part the unified approach to solving inverse Sturm-Liouville problems is presented, applied to four different problems: the inverse problem on a finite interval, the inverse problem on a half-line, the inverse quantum scattering problem on a half-line and the inverse scattering problem on a line. The exposition includes numerical examples and illustrations.

The author hopes that this book will contribute to the further development of this magnificent and important piece of modern mathematics.

Part I
Typical Problem Statements

We give a brief description of the direct and inverse spectral problems considered in this book as well as some necessary preliminaries on Sturm-Liouville equations, such as the spectral parameter power series representations for solutions and Liouville's transformation.

Preliminaries on Sturm-Liouville Equations

2

2.1 Notation, Abel's Formula

We will consider the second-order linear ordinary differential equation

$$- y'' + q(x)y = \lambda y \tag{2.1}$$

on a finite or infinite interval. This equation is called the *Sturm-Liouville equation*, or often the one-dimensional *Schrödinger equation*. Conditions satisfied by the coefficient $q(x)$, frequently called the *potential*, will be specified along the way, depending on the problem under consideration. The complex number λ is called the *spectral parameter*. Very often it will be convenient to work with its square root $\rho := \sqrt{\lambda}$, which is usually chosen so that $\operatorname{Im} \rho \geq 0$.

Assume that q is a complex-valued function belonging to $L_1(a, b)$. Then for every fixed λ, Eq. (2.1) possesses two linearly independent solutions y_1 and y_2. It is easy to see that their Wronskian is constant, $W[y_1, y_2] := y_1 y_2' - y_1' y_2 = \text{const}$. This is a corollary of the Liouville-Ostrogradski formula (which in the case of linear second-order equations was discovered by N. H. Abel) and the fact that one can find a general solution of (2.1) if one nontrivial particular solution is known. Indeed, a second linearly independent solution is provided the *Abel formula*

$$y_2(x) = y_1(x) \int_{x_0}^{x} \frac{dt}{y_1^2(t)}, \quad x_0 \in [a, b] \tag{2.2}$$

whenever the integration makes sense.

© The Editor(s) (if applicable) and The Author(s), under exclusive licence to Springer Nature Switzerland AG 2020
V. V. Kravchenko, *Direct and Inverse Sturm-Liouville Problems*,
Frontiers in Mathematics, https://doi.org/10.1007/978-3-030-47849-0_2

2.2   Spectral Parameter Power Series Representations

Let q be a complex-valued function of the real variable $x \in [a, b]$, let $q \in L_1(a, b)$ and let λ be an arbitrary complex number. The interval (a, b) is assumed to be finite. Suppose there exists a solution f of the equation

$$f'' - q(x)f = 0$$

on (a, b) such that $f(x) \neq 0$ for all $x \in [a, b]$. In general, f is complex-valued. Consider the following two families of functions defined recursively as

$$X^{(0)}(x) \equiv 1, \qquad X^{(n)}(x) = n \int_{x_0}^{x} X^{(n-1)}(s) \left(f^2(s)\right)^{(-1)^n} ds,$$

$$x_0 \in [a, b], \quad n = 1, 2, \ldots \qquad (2.3)$$

and

$$\widetilde{X}^{(0)} \equiv 1, \qquad \widetilde{X}^{(n)}(x) = n \int_{x_0}^{x} \widetilde{X}^{(n-1)}(s) \left(f^2(s)\right)^{(-1)^{n-1}} ds,$$

$$x_0 \in [a, b], \quad n = 1, 2, \ldots. \qquad (2.4)$$

The functions $\{\varphi_k\}_{k=0}^{\infty}$ defined by

$$\varphi_k(x) = \begin{cases} f(x)X^{(k)}(x), & k \text{ odd}, \\ f(x)\widetilde{X}^{(k)}(x), & k \text{ even} \end{cases} \qquad (2.5)$$

will be called *formal powers* associated to the Sturm-Liouville equation

$$-y'' + q(x)y = \lambda y, \quad a < x < b. \qquad (2.6)$$

The system of functions (2.5) is closely related to the notion of the L-basis introduced and studied in [65]. Here L stands for a linear ordinary differential operator.

Together with the system of functions (2.5), we introduce the functions $\{\psi_k\}_{k=0}^{\infty}$ using the "second half" of the recursive integrals (2.3) and (2.4):

$$\psi_k(x) = \begin{cases} \dfrac{\widetilde{X}^{(k)}(x)}{f(x)}, & k \text{ odd}, \\[3mm] \dfrac{X^{(k)}(x)}{f(x)}, & k \text{ even}. \end{cases} \qquad (2.7)$$

The following result, obtained in [100] (for additional details and simpler proof see [101] and [108]), provides a general solution of Eq. (2.6) in the form of a *spectral parameter power series* (SPPS). It was given in [100, 101] and [108] for continuous coefficients, and then in [32] it was extended to the case $q \in L_1(a, b)$.

Theorem 2.1 (SPPS Representation) *Let q be a complex valued function of the real variable $x \in [a, b]$, let $q \in L_1(a, b)$ and let λ be an arbitrary complex number. Suppose there exists a solution f of the equation*

$$f'' - q(x)f = 0 \tag{2.8}$$

on (a, b) such that $f(x) \neq 0$ for all $x \in [a, b]$. Then the general solution y of Eq. (2.6) has the form

$$y = c_1 y_1 + c_2 y_2, \tag{2.9}$$

where c_1 and c_2 are arbitrary complex constants,

$$y_1 = \sum_{k=0}^{\infty} \frac{(-\lambda)^k}{(2k)!} \varphi_{2k} \quad and \quad y_2 = \sum_{k=0}^{\infty} \frac{(-\lambda)^k}{(2k+1)!} \varphi_{2k+1}. \tag{2.10}$$

Both series converge uniformly on $[a, b]$ and the series of the first derivatives, which have the form

$$y_1' = f' + \sum_{k=1}^{\infty} \frac{(-\lambda)^k}{(2k)!} \left(\frac{f'}{f} \varphi_{2k} + 2k \, \psi_{2k-1} \right) \quad and$$

$$y_2' = \sum_{k=0}^{\infty} \frac{(-\lambda)^k}{(2k+1)!} \left(\frac{f'}{f} \varphi_{2k+1} + (2k+1) \, \psi_{2k} \right) \tag{2.11}$$

converge to y_1' and y_2', respectively, in the space $L_1(a, b)$.

   For every fixed x, the series (2.10) converge uniformly on any compact subset of the complex ρ-plane and hence for every fixed x, y_1 and y_2 are entire functions of the complex variable ρ (and, of course, of λ).

Remark 2.1 It is easy to see that by definition the solutions y_1 and y_2 in (2.10) satisfy the initial conditions

$$y_1(x_0) = f(x_0), \qquad y_1'(x_0) = f'(x_0),$$
$$y_2(x_0) = 0, \qquad y_2'(x_0) = 1/f(x_0).$$

Remark 2.2 In the case $\lambda = 0$, the solutions (2.10) become $y_1 = f$ and $y_2 = f \int\limits_{x_0}^{x} \frac{ds}{f^2(s)}$.
The expression for y_2 is the Abel formula (2.2).

Remark 2.3 Theorem 2.1 remains valid for infinite intervals, with the series uniformly convergent on any finite subinterval.

Remark 2.4 One of the functions f^2 or $1/(f^2)$ may not be continuous on $[a, b]$, and yet y_1 or y_2 may make sense. For example, in the case of the Bessel equation $(xy')' - \frac{1}{x}y = -\lambda xy$, we can choose $f(x) = x/2$. Then $1/(f^2) \notin C[0, 1]$. Nevertheless, all the integrals in (2.4) exist and y_1 coincides with the nonsingular Bessel function $J_1(\sqrt{\lambda}x)$, while y_2 is a singular solution of the Bessel equation.

Remark 2.5 The procedure for construction of solutions described in Theorem 2.1 works not only when a solution is available for $\lambda = 0$, but in fact when a solution of the equation

$$- y_0'' + q(x)y_0 = \lambda_0 y_0 \tag{2.12}$$

is known for some fixed λ_0. The solutions (2.10) now take the form

$$y_1 = y_0 \sum_{k=0}^{\infty} (-1)^k (\lambda - \lambda_0)^k \frac{\widetilde{X}^{(2k)}}{(2k)!} \quad \text{and} \quad y_2 = y_0 \sum_{k=0}^{\infty} (-1)^k (\lambda - \lambda_0)^k \frac{X^{(2k+1)}}{(2k+1)!}.$$

This can be easily verified by writing (2.6) as

$$(L - \lambda_0)\, y = (\lambda - \lambda_0)\, y, \quad L := -\frac{d^2}{dx^2} + q(x).$$

The operator on the left-hand side can be factorized exactly as in the proof of the theorem, and the same reasoning carries through.

The possibility to represent solutions of the Sturm-Liouville equation in such form is by no means a novelty, though it is not a widely used tool. We mention [29, 62, 125, Sect. 10] and [97]. To our best knowledge, it was applied for solving spectral problems for the first time in [108]. The reason of this underuse of the SPPS lies in the form in which the expansion coefficients were sought. Indeed, in previous works the calculation of coefficients was carried out in terms of successive integrals with the kernels given by iterated Green functions (see [29, Sect. 10]). This makes any computation based on such representation difficult and less practical; even proofs of the most basic results like, e.g., the uniform convergence of the SPPS for any value of $\lambda \in \mathbb{C}$ (established in Theorem 2.1) are not an easy task. For example, in [29, p. 16] the parameter λ is assumed to be small and no proof of convergence is given.

Calculating the expansion coefficients in (2.10) as indicated in (2.3), (2.4) is relatively simple and straightforward; this is why the estimation of the rate of convergence of the series (2.10) presents no difficulty, see [32, 108]. Moreover, in [38] a discrete analogue of Theorem 2.1 was established and the discrete analogues of the series (2.10) turned out to be finite sums.

Another crucial feature of the introduced representation of the expansion coefficients in (2.10) is that not only these coefficients (denoted by φ_k in (2.5)) are required for solving various spectral problems related to the Sturm-Liouville equation. Indeed, while the functions $\widetilde{X}^{(2k+1)}$ and $X^{(2k)}$, $k = 0, 1, 2, \ldots$ do not participate explicitly in the representation (2.10), together with the functions φ_k they appear in the representation (2.11) of the derivatives of the solutions and therefore also in characteristic equations corresponding to the spectral problems (see Sect. 11.1 below).

Remark 2.6 It is worth mentioning that in the case of a real-valued coefficient q the existence and construction of the required f presents no difficulty. Let q be real-valued on $[a, b]$. Then Eq. (2.8) has two linearly independent regular solutions v_1 and v_2 whose zeros alternate. Thus one may choose $f = v_1 + iv_2$. Moreover, for the construction of v_1 and v_2 one can in fact use the same SPPS method [108].

In general, the existence of a nonvanishing solution for complex-valued q was proved in [108, Remark 5] (see also [37]). Moreover, even when f has zeros, the formal powers are still well defined and can be constructed following, e.g., a procedure given in [114]. Thus, in what follows we will not restrict ourselves to the requirement $f \neq 0$.

The SPPS representation was obtained in [101, 108] for a general Sturm-Liouville equation of the form (2.13), for pencils of Sturm-Liouville operators in [114, 121], for Bessel-type singular Sturm-Liouville equations in [47, 48], for higher-order Sturm-Liouville equations in [87, 110], and for Dirac-type systems of equations in [77]. It was used in dozens of publications as a convenient tool for studying different models in mathematics and mathematical physics (see [19–27, 36, 45, 46, 48, 63, 78, 79, 88, 93–96, 109, 122, 129, 134, 143–148, 175, 176]).

Besides finding solutions of Eq. (2.1) satisfying additional boundary or initial conditions, it is often necessary to solve spectral problems, also called Sturm-Liouville problems. Their variety is enormous. We will only consider some of the most frequently encountered and canonical such problems.

2.3   Liouville's Transformation

Linear differential equations of a more general form can be reduced to the form (2.1). Under certain conditions on the coefficients, the Liouville transformation can be used for this purpose.

Consider the Sturm-Liouville differential equation

$$(p(t)v')' - q(t)v = -\lambda r(t)v, \quad t \in [\alpha, \beta], \tag{2.13}$$

with $[\alpha, \beta]$ a finite interval. Let $q : [\alpha, \beta] \to \mathbb{C}$, p and $r : [\alpha, \beta] \to \mathbb{R}$ be such that $q \in C[\alpha, \beta]$, $p, p', r, r' \in AC[\alpha, \beta]$, and $p(t) > 0$, $r(t) > 0$ for all $t \in [\alpha, \beta]$. Here $AC[\alpha, \beta]$ denotes the space of all absolutely continuous functions on $[\alpha, \beta]$. Define the mapping $l : [\alpha, \beta] \to [0, b]$ by

$$l(t) := \int_{\alpha}^{t} \{r(s)/p(s)\}^{1/2} \, ds, \tag{2.14}$$

where $b = \int_{\alpha}^{\beta} \{r(s)/p(s)\}^{1/2} \, ds$. Denote $\theta(t) := (p(t)r(t))^{1/4}$. Then (2.13) is related to the one-dimensional Schrödinger differential equation

$$u'' - Q(x)u = -\lambda u, \quad x \in [0, b], \tag{2.15}$$

by the Liouville transformation of the variables t and v into x and u, defined by (see, e.g., [64, 177])

$$x = l(t), \quad u(x) := \theta(t)v(t) \quad \text{for all } t \in [\alpha, \beta]$$

and the coefficient Q is given by the formula

$$\begin{aligned}
Q(x) &= \frac{q(t)}{r(t)} + \frac{p(t)}{4r(t)} \left[\left(\frac{p'(t)}{p(t)} + \frac{r'(t)}{r(t)} \right)' + \frac{3}{4} \left(\frac{p'(t)}{p(t)} \right)^2 + \frac{1}{2} \frac{p'(t)}{p(t)} \frac{r'(t)}{r(t)} - \frac{1}{4} \left(\frac{r'(t)}{r(t)} \right)^2 \right] \\
&= \frac{q(t)}{r(t)} - \frac{\theta(t)}{r(t)} \left[p(t) \left(\frac{1}{\theta(t)} \right)' \right]' \quad \text{for all } t \in [\alpha, \beta].
\end{aligned} \tag{2.16}$$

The Liouville transformation can be regarded as an operator $\mathbf{L} : C[\alpha, \beta] \to C[0, b]$ acting according to the rule

$$u(x) = \mathbf{L}[v(t)] = \theta(l^{-1}(x))v(l^{-1}(x)).$$

Let us introduce the following notations for the differential expressions:

$$\mathbf{B} = -\frac{d^2}{dx^2} + Q(x) \quad \text{and} \quad \mathbf{C} = -\frac{1}{r(t)} \left(\frac{d}{dt} \left(p(t)\frac{d}{dt} \right) - q(t) \right).$$

The following proposition summarizes the main properties of the operator **L**.

Proposition 2.1

1. *The inverse operator is given by* $v(t) = \mathbf{L}^{-1}[u(x)] = \frac{1}{\theta(t)}u(l(t))$.
2. *The uniform norms of the operators* **L** *and* $\mathbf{L}^{-1}$ *are equal to* $\|\mathbf{L}\| = \max_{t\in[\alpha,\beta]} |\theta(t)|$ *and* $\|\mathbf{L}^{-1}\| = \max_{t\in[\alpha,\beta]} |1/\theta(t)|$, *respectively.*
3. *The operator equality*

$$\mathbf{BL} = \mathbf{LC}$$

holds on $C^2(\alpha, \beta) \cap C[\alpha, \beta]$.

It is worth mentioning that the Liouville transformation maps one-to-one the formal powers associated with Eq. (2.13) to the formal powers associated with Eq. (2.15). The interested reader can find more on this subject in [105] and [118].

Direct and Inverse Sturm-Liouville Problems on Finite Intervals

3

3.1 Direct Problems

Consider Eq. (2.1) on the interval $(0, \pi)$. If originally the equation is considered on another finite interval (a, b), then by a simple change of the independent variable $t = \frac{x-a}{b-a}\pi$ it can always be transferred to $(0, \pi)$. Let q be a real-valued function, $q \in L_2(0, \pi)$ and h, H be real numbers. Together with the equation

$$- y'' + q(x)y = \lambda y, \quad 0 < x < \pi \tag{3.1}$$

consider the homogeneous boundary conditions

$$y'(0) - hy(0) = y'(\pi) + Hy(\pi) = 0. \tag{3.2}$$

There exists an infinite sequence of real numbers $\{\lambda_n\}_{n=0}^{\infty}$ such that $\lambda_n < \lambda_m$ if $n < m$, $\lambda_n \to +\infty$ when $n \to \infty$, and for every λ_n the equation

$$-y_n'' + q(x)y_n = \lambda_n y_n, \quad 0 < x < \pi$$

admits a nontrivial solution y_n satisfying the conditions (3.2).

These values λ_n of the spectral parameter are called *eigenvalues of the Sturm-Liouville problem* (3.1), (3.2) while the corresponding solutions y_n are called *eigenfunctions of the Sturm-Liouville problem* (3.1), (3.2). The multiplicity of every eigenvalue of this problem is equal to one, which means that there exists exactly one, up to a multiplicative constant, corresponding eigenfunction. It is often convenient to normalize the eigenfunctions. For

© The Editor(s) (if applicable) and The Author(s), under exclusive licence to Springer Nature Switzerland AG 2020
V. V. Kravchenko, *Direct and Inverse Sturm-Liouville Problems*,
Frontiers in Mathematics, https://doi.org/10.1007/978-3-030-47849-0_3

this purpose, we let $\varphi(\rho, x)$ denote solution of the Cauchy problem

$$-\varphi'' + q(x)\varphi = \lambda\varphi, \tag{3.3}$$

$$\varphi(0) = 1, \quad \varphi'(0) = h \tag{3.4}$$

where $\lambda = \rho^2$. This function exists and is unique. For all λ it satisfies the first of the conditions (3.2). Clearly, when $\lambda = \lambda_n = \rho_n^2$ is an eigenvalue, the function $\varphi(\rho_n, x)$ is an eigenfunction belonging to the eigenvalue λ_n. It is said to be *normalized at the origin*.

To solve the *direct* Sturm-Liouville problem means to find its complete spectral data $\left\{\lambda_n, \varphi(\rho_n, x)\right\}_{n=0}^{\infty}$.

It is often necessary to compute the so-called *norming constants* or *weight numbers*, namely, the squares of the L_2-norms of the normalized eigenfunctions,

$$\alpha_n := \int_0^{\pi} \varphi^2(\rho_n, x)dx.$$

The following asymptotic relations are valid for the sequences $\{\lambda_n\}_{n=0}^{\infty}$ and $\{\alpha_n\}_{n=0}^{\infty}$:

$$\rho_n := \sqrt{\lambda_n} = n + \frac{\omega}{\pi n} + \frac{k_n}{n}, \quad \alpha_n = \frac{\pi}{2} + \frac{K_n}{n}, \quad \{k_n\}, \{K_n\} \in \ell_2. \tag{3.5}$$

The number ω appearing in the asymptotics of ρ_n is given by the formula

$$\omega = h + H + \frac{1}{2}\int_0^{\pi} q(t)\, dt. \tag{3.6}$$

It is well known that the eigenfunctions belonging to different eigenvalues are orthogonal:

$$\int_0^{\pi} \varphi(\rho_m, x)\varphi(\rho_n, x)dx = 0 \quad \text{for } n \neq m,$$

and for the arbitrary functions $u \in L_2(0, \pi)$ and $v \in L_2(0, \pi)$ the *Parseval identity* is valid

$$\int_0^{\pi} u(x)v(x)dx = \sum_{n=0}^{\infty} \frac{1}{\alpha_n^2} \int_0^{\pi} u(x)\varphi(\rho_n, x)dx \int_0^{\pi} v(x)\varphi(\rho_n, x)dx.$$

The theory of Sturm-Liouville problems is presented in numerous books (see, e.g., [10, 128, 174]) and is a part of basic courses on partial differential equations and equations of mathematical physics, so that we restrict ourselves to these few definitions and results, referring the reader to the available bibliography.

3.2   Inverse Problems

The inverse Sturm-Liouville problem on a finite interval is formulated as follows. Given two sequences of real numbers $\{\lambda_n\}_{n=0}^{\infty}$ and $\{\alpha_n\}_{n=0}^{\infty}$ such that $\lambda_n < \lambda_m$ for $n < m$, $\alpha_n > 0$, and the relations (3.5) are valid. Find the real-valued potential $q(x)$ and the real numbers h and H such that $\{\lambda_n\}_{n=0}^{\infty}$ is the spectrum of the Sturm-Liouville problem (3.1), (3.2) and α_n, $n = 0, 1, \ldots$ are the corresponding norming constants. This problem possesses a unique solution $q \in L_2(0, \pi)$, $h, H \in \mathbb{R}$ (see, e.g., [127, 133, 172]).

Often another version of the inverse Sturm-Liouville problem on a finite interval is considered, in which instead of the sequence of norming constants another sequence of eigenvalues $\{\mu_n\}_{n=0}^{\infty}$ is given, corresponding to a Sturm-Liouville problem with the same Eq. (3.1), but with other boundary conditions, for example, $y'(0) - hy(0) = y'(\pi) + H_1 y(\pi) = 0$ with $H_1 \neq H$. The problem of finding the potential $q \in L_2(0, \pi)$ and the coefficients h, H, H_1 in the boundary conditions is known as the two-spectra inverse Sturm-Liouville problem. The two stated inverse Sturm-Liouville problems are equivalent. The knowledge of a second spectrum allows one to find the sequence of the norming constants and vice versa (see details in [127, 172]). We will consider the first of the problems: the recovering of the potential by one spectrum and norming constants.

Quite a few numerical methods have been developed for solving both direct and inverse Sturm-Liouville problems on finite intervals. The main approach for solving direct Sturm-Liouville problems reduces to the shooting method, which consists in solving the Cauchy problem (3.3), (3.4) at each step (making a "shot") by any available method and finding the closest possible values of λ that make the expression $\varphi'(\pi) + H\varphi(\pi)$ equal to zero. The differences between known methods of solution of the direct Sturm-Liouville problems thus reside in the methods employed for solving the Cauchy problem. Recommended references on this subject are the book [142] and the work [124]. In Part III of this text, two recently introduced methods for solving direct Sturm-Liouville problems will be presented. Both of them rely on the notion and properties of the transmutation operator. An advantage of both methods is that they yield one a characteristic function of the problem in an analytical form which then can be used for further work. For example, in many situations not the eigenvalues themselves may be of interest, but rather their first or even second derivatives with respect to certain physically relevant parameters, like in study of dispersion coefficients in models of wave propagation through nonhomogeneous optical fibers [48]. Both methods offer such possibilities to the difference of purely numerical approaches. The first method, based on the spectral parameter power series (SPPS) from Sect. 2.2, became since the first publications in which it was presented [100, 101, 108] a widely used tool for solving Sturm-Liouville problems (see references in Sect. 2.2) due to its remarkable simplicity and the fact that it allows to write down explicitly characteristic functions of the problems. The second was introduced in [106] and can be called the Neumann series of Bessel functions (NSBF) method because this is precisely the type of the series representations derived for the solutions of the Sturm-Liouville equation. We present the NSBF representations in Chap. 9.

The SPPS method is convenient when the solution is sought in a not too large domain in the complex plane of the spectral parameter, while the NSBF method allows one to compute very large sets of eigendata, since the NSBF representation of the solution admits truncation estimates independent of Re ρ.

Various inverse Sturm-Liouville problems on finite intervals found applications in numerous fields (see, e.g., [50, 75, 166, 167]). Very often in practice the potential should be recovered from relatively few eigendata.

Several approaches have been developed for the numerical solution of inverse Sturm-Liouville problems on finite intervals (see, e.g., [14–16, 35, 69, 80, 83, 96, 130, 135, 137, 149, 151, 154, 156]). All competitive approaches are iterative. In Chap. 13 we present a simple and direct method which reduces the inverse Sturm-Liouville problem to a system of linear algebraic equations. It was introduced in [103]. The method is related to the NSBF representation and to the famous Gelfand-Levitan equation from the theory of inverse Sturm-Liouville problems.

Direct and Inverse Sturm-Liouville Problems on a Half-Line

Consider Eq. (2.1) on a half-line:

$$-y'' + q(x)y = \lambda y, \quad x > 0, \tag{4.1}$$

where $q(x)$ is a real-valued function satisfying the condition

$$\int_0^\infty (1+x)\,|q(x)|\,dx < \infty. \tag{4.2}$$

Roughly speaking, this condition means that the potential decays rather rapidly at infinity and is integrable on any finite interval. In physics such potentials are called *short-range potentials*. Consider an initial condition

$$y'(0) - hy(0) = 0, \quad h \in \mathbb{R}. \tag{4.3}$$

Under the conditions imposed, the Sturm-Liouville problem (4.1), (4.3) has at most finitely many simple eigenvalues; they are the values of the spectral parameter λ for which the problem (4.1), (4.3) admits nontrivial solutions belonging to $L_2(0, \infty)$. The eigenvalues, if they exist, are all negative. Together with the eigenvalues, the corresponding norming constants

$$\alpha_n = \left(\int_0^\infty \varphi^2(\rho_n, x)dx \right)^{-1} \tag{4.4}$$

V. V. Kravchenko, *Direct and Inverse Sturm-Liouville Problems*, Frontiers in Mathematics, https://doi.org/10.1007/978-3-030-47849-0_4

are sought, where ρ_n is the square root of an eigenvalue and $\varphi(\rho_n, x)$ is a solution of the Cauchy problem (3.3), (3.4) with $\lambda = \rho_n^2$, hence $\varphi(\rho_n, x)$ is the associated normalized eigenfunction.

The spectrum of the problem includes a continuous part which is the positive half-line $\lambda > 0$ on which the *spectral density function* $\sigma(\lambda)$ or, even more, its derivative $V(\lambda) := \sigma'(\lambda)$ are of great interest. The continuous spectrum of the problem is the set of those values of $\lambda - \rho^2$ for which the corresponding solution $\varphi(\rho, x)$ of the Cauchy problem (3.3), (3.4) is bounded, yet does not belong to $L_2(0, \infty)$ (it oscillates, but does not decay at infinity, at least not sufficiently fast). The spectral density function is absolutely continuous, which implies that it is differentiable almost everywhere. Moreover, $V(\lambda) > 0$, $\lambda > 0$.

Thus, the set of spectral data of the Sturm-Liouville problem on a half-line consists of a finite (if any) set of eigenvalues and norming constants $\{\lambda_k = \rho_k^2 < 0, \alpha_k\}_{k=1}^m$, and the function $V(\lambda) > 0$, $\lambda > 0$ (related to the spectral density function as $\sigma'(\lambda) = V(\lambda)$, $\lambda > 0$).

All the spectral data are present in one of the central facts of the spectral theory, the *spectral expansion theorem*. Following [172] we denote by W_1 the set of functions $f(x)$, $x \geq 0$ such that $f^{(j)}(x) \in AC[0, T]$ (absolutely continuous) $j = 0, 1$ for any $T > 0$ fixed, and $f^{(j)}(x) \in L_1(0, \infty)$, $j = 0, 1$. For all $f(x) \in W_1$ the following spectral expansion theorem is valid.

Theorem 4.1 (See, e.g., [69, Th. 2.1.8], [172, Section 2.1.3]) *Let $f(x) \in W_1$. Then*

$$f(x) = \int_0^\infty \varphi(\rho, x) F(\lambda) V(\lambda) d\lambda + \sum_{n=1}^N \varphi(\rho_n, x) F(\lambda_n) \alpha_n, \qquad (4.5)$$

uniformly with respect to x, where

$$F(\lambda) = \int_0^\infty \varphi(\rho, t) f(t) dt.$$

To solve the direct Sturm-Liouville problem on a half-line means to find the whole set of the spectral data: the corresponding to the discrete spectrum $\{\lambda_k, \alpha_k\}_{k=\overline{1,m}}$ and the corresponding to the continuous spectrum function $V(\lambda) > 0$, $\lambda > 0$.

While the computation of a finite set of the discrete spectral data is quite attainable and a good number of publications are dedicated to this subject, the computation of $V(\lambda)$ has always been considered as a much more difficult task. Few works exist in which the authors propose algorithms for numerical calculation of $\sigma(\lambda)$ [71, 72, 142]. In the last section of the book [142] the author explains the difficulty of this problem. The essential challenge consists in finding a good way of approximating the problem on a half-line by problems on finite intervals. On each finite interval, may it be large, the corresponding Sturm-Liouville problem will have a discrete spectrum only. How the function $V(\lambda)$ (or $\sigma(\lambda)$) is related to the distribution of the eigenvalues of the problems on finite intervals is not an easy question. The interested reader can find some respective information in [142, Sect. 11.10], where it is also stated that the computation is expensive, requiring hundreds of thousands of eigenvalues and eigenfunction norms to be computed for obtaining the values of $\sigma(\lambda)$ in several points and with a modest accuracy.

One of the results presented in Part III is a method of computation of $V(\lambda)$ based on a completely different idea and allowing one to reduce the computation of $V(\lambda)$ on a half-line $\lambda > 0$ to computation of a quite simple expression on a finite interval. Numerical tests show that $V(\lambda)$ is obtained within a fraction of second on a large interval of the variable λ and with a remarkable accuracy.

In order to proceed with the study of the Sturm-Liouville problem on a half-line let us introduce the concept of the Jost solution.

It is well known (see, e.g., [69], [172, Theorem 2.4.1]) that under the assumption (4.2) Eq. (4.1) possesses the unique so-called *Jost solution* $y = e(\rho, x)$ such that for $v = 0, 1$,

$$e^{(v)}(\rho, x) = (i\rho)^v e^{i\rho x} (1 + o(1)), \quad x \to \infty, \tag{4.6}$$

uniformly in $\overline{\Omega_+}$, where $\Omega_+ := \{\rho \in \mathbb{C} : \operatorname{Im} \rho > 0\}$. For every fixed $\rho \in \Omega_+$, the function $e(\rho, x)$ belongs to $L_2(0, \infty)$. When $\operatorname{Im} \rho = 0$, it is only bounded.

Moreover,

$$e^{(v)}(\rho, x) = (i\rho)^v e^{i\rho x} \left(1 + \frac{\omega(x)}{i\rho} + o\left(\frac{1}{\rho}\right)\right), \quad |\rho| \to \infty, \tag{4.7}$$

$$\omega(x) := -\frac{1}{2} \int_x^\infty q(t)dt$$

uniformly with respect to $x \geq 0$ [172, p. 129].

In terms of the Jost solution all the spectral data can be expressed as follows (see, e.g., [172, pp. 167, 168]). Denote

$$\Delta(\rho) := e'(\rho, 0) - he(\rho, 0)$$

and

$$\Delta_1(\rho) := \frac{d}{d\lambda}\Delta(\rho) \quad (\lambda = \rho^2).$$

Then the eigenvalues can be computed as squares of zeros of $\Delta(\rho)$, thus $\Delta(\rho_k) = 0$, $k = \overline{1, \ldots, m}$. The corresponding norming constants satisfy the equality

$$\alpha_k = \frac{e(\rho_k, 0)}{\Delta_1(\rho_k)}, \tag{4.8}$$

and, finally,

$$V(\lambda) = \frac{\rho}{\pi \, |\Delta(\rho)|^2}, \quad \rho > 0. \tag{4.9}$$

Thus, having computed

$$e(\rho, x), \; e'(\rho, x), \; \frac{\partial e(\rho, x)}{\partial \rho} \; \text{ and } \; \frac{\partial e'(\rho, x)}{\partial \rho}$$

one can solve the Sturm-Liouville problem on a half-line. In Part II we derive convenient series representations for these four functions which indeed make such computation possible.

The inverse Sturm-Liouville problem on a half-line is stated as follows. Given the spectral data $\{\lambda_k, \alpha_k\}_{k=1}^m$ and $V(\lambda) > 0$, $\lambda > 0$. Find a function $q(x)$, $x > 0$, satisfying (4.2) and a constant $h \in \mathbb{R}$ such that the problem (4.1), (4.3) possess the given spectral data.

For additional information on this problem, we refer to [127] and [172]. Applications of direct and inverse Sturm-Liouville problems on a half-line to solving direct and inverse problems for partial differential equations can be found, e.g., in [123] and [152]. In Chap. 14 a simple and direct method for solving the inverse Sturm-Liouville problem on a half-line is presented. It was introduced in [58] and allows one to reduce the problem to a system of linear algebraic equations.

Quantum Scattering Problem on the Half-Line

5

Consider Eq. (4.1), where the potential q satisfies the condition (4.2). Let $e(\rho, x)$ be the corresponding Jost solution. The function $F(\rho) := e(\rho, 0)$ is called the *Jost function* and the quotient

$$S(\rho) := \frac{F(-\rho)}{F(\rho)} \tag{5.1}$$

is traditionally called the scattering matrix, or simply *S-matrix* (see, e.g., [51]). Notice that due to (4.7) we have that

$$F(\rho) = 1 + \frac{\omega(x)}{i\rho} + o\left(\frac{1}{\rho}\right), \quad |\rho| \to \infty. \tag{5.2}$$

Instead of the initial condition (4.3), consider the condition

$$y(0) = 0. \tag{5.3}$$

The potential q may admit a finite number of bound states (eigenfunctions), the physically meaningful solutions of (4.1), (5.3) that are square integrable on $(0, \infty)$. They may exist for certain negative values of the parameter λ, called eigenvalues. Usually, the *scattering data* are introduced as

$$\left\{ S(\rho), \ \rho > 0; \ \left\{ \rho_j^2, c_j \right\}_{j=1}^{N} \right\}, \tag{5.4}$$

V. V. Kravchenko, *Direct and Inverse Sturm-Liouville Problems*, Frontiers in Mathematics, https://doi.org/10.1007/978-3-030-47849-0_5

where ρ_j^2 are the eigenvalues of the problem and c_j are the corresponding norming constants [127],

$$c_j := \frac{1}{\int_0^\infty \psi^2(\rho_j, x)dx} = -\frac{2\rho_j e'(\rho_j, 0)}{F'(\rho_j)}. \tag{5.5}$$

Here $\psi(\rho, x)$ denotes the solution of (4.1) satisfying the initial conditions

$$\psi(\rho, 0) = 0, \quad \psi'(\rho, 0) = 1.$$

The direct quantum scattering problem consists in finding the scattering data (5.4) when the potential q satisfying (4.2) is given. The inverse scattering problem requires finding q when the scattering data are given [51].

The solution of the direct scattering problem reduces in fact to finding the Jost solution and its derivative. Then the eigenvalues (if they exist) correspond to zeros of the Jost function $F(\rho)$, $\operatorname{Im} \rho > 0$. Indeed, if ρ^*, with $\operatorname{Im} \rho^* > 0$, is such that $F(\rho^*) = 0$, then the corresponding Jost solution $e(\rho^*, x)$ satisfies Eq. (4.1), the condition (5.3), and belongs to $L_2(0, \infty)$ and hence it is a bound state. The norming constants are computed by (5.5), and $S(\rho)$ is obtained from (5.1).

For more information on direct and inverse quantum scattering problems we refer the reader to [50] and [51]. In Chap. 12 a practical method for solving the direct problem based on a convenient series representation for the Jost solution and its derivative from Chap. 10 is presented. In Chap. 15 a simple and direct method for solving the inverse quantum scattering problem is derived. It was introduced in [85], and similarly to the methods for solving the inverse Sturm-Liouville problem on a finite interval (Chap. 13) and the inverse Sturm-Liouville problem on a half-line (Chap. 14), it reduces the inverse problem to a system of linear algebraic equations.

Scattering Problem on the Line

6

Consider now the one-dimensional Schrödinger equation on the whole real line:

$$- y'' + q(x)y = \lambda y, \quad x \in (-\infty, \infty), \tag{6.1}$$

where $q(x)$ is a real-valued function defined on $(-\infty, \infty)$ and satisfies the condition

$$\int_{-\infty}^{\infty} (1 + |x|) \, |q(x)| \, dx < \infty. \tag{6.2}$$

Besides the Jost solution at plus infinity, let us introduce the Jost solution at minus infinity, defined by the asymptotic relations,

$$g^{(\nu)}(\rho, x) = (-i\rho)^{\nu} e^{-i\rho x} (1 + o(1)), \quad x \to -\infty, \quad \nu = 0, 1,$$

uniformly in $\overline{\Omega_+}$.

When $\rho \in \mathbb{R}$ we have that

$$e(-\rho, x) = \overline{e(\rho, x)}, \quad g(-\rho, x) = \overline{g(\rho, x)}$$

and both sets of solutions

$$\left\{ e(\rho, x), \overline{e(\rho, x)} \right\} \quad \text{and} \quad \left\{ g(\rho, x), \overline{g(\rho, x)} \right\}$$

V. V. Kravchenko, *Direct and Inverse Sturm-Liouville Problems*,
Frontiers in Mathematics, https://doi.org/10.1007/978-3-030-47849-0_6

are fundamental systems of solutions of (6.1) for $\rho \neq 0$. In particular, $g(\rho, x)$ and $\overline{g(\rho, x)}$ can be represented as linear combinations of the fundamental system $\left\{ e(\rho, x), \overline{e(\rho, x)} \right\}$:

$$g(\rho, x) = a(\rho)\overline{e(\rho, x)} + b(\rho)e(\rho, x),$$

$$\overline{g(\rho, x)} = \overline{a(\rho)}e(\rho, x) + \overline{b(\rho)e(\rho, x)}.$$

The quantities $a(\rho)$ and $b(\rho)$ are called *scattering amplitudes* and can be calculated by the formulas

$$a(\rho) = -\frac{1}{2i\rho} W\left[e(\rho, x), g(\rho, x) \right], \quad \rho \in \overline{\Omega_+},$$

$$b(\rho) = \frac{1}{2i\rho} W\left[e(\rho, x), g(-\rho, x) \right], \quad \rho \in \mathbb{R}.$$

Notice that the second expression is well defined for real values of ρ only, because when $\text{Im}\,\rho > 0$, a solution of (6.1) behaving as $e^{-i\rho x}$ when $x \to -\infty$ is not unique.

The scattering amplitudes are closely related to the reflection and transmission coefficients. Imagine that a plane wave of unit amplitude, asymptotically similar to $e^{-i\rho x}$ when $x \to \infty$, $\rho \in \mathbb{R}$, comes from the right (from the plus infinity), as it is schematically depicted on Fig. 6.1. It interacts with a medium characterized by the potential q. A part

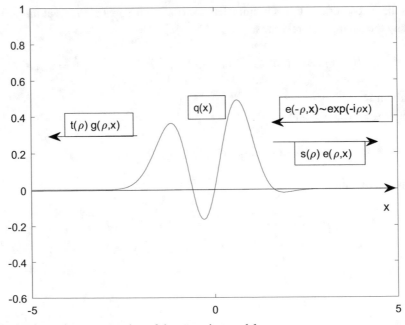

Fig. 6.1 A schematic representation of the scattering model

of it is transmitted retaining the same phase but with a new amplitude depending on ρ: $t(\rho)$, while another part is reflected, changing its phase to the opposite one and with an amplitude $s(\rho)$. These amplitudes of the transmitted part of a unit plane wave and of its reflected part are called the *transmission* and *reflection coefficients*, respectively. From the figure it is seen that

$$\overline{e(\rho, x)} = t(\rho)g(\rho, x) - s(\rho)e(\rho, x).$$

Thus, $t(\rho) = 1/a(\rho)$ and $s(\rho) = b(\rho)/a(\rho)$.

We considered the situation of a wave travelling from the right to the left. Analogously, a wave travelling in the opposite direction can be considered, and hence two reflection coefficients (the right and the left) can be introduced instead. They have the form

$$s^{\pm}(\rho) = \mp \frac{b(\mp\rho)}{a(\rho)}.$$

To solve the *scattering problem* means to find a finite set (if it is not empty) of negative eigenvalues $\lambda_n = \rho_n^2 = (i\tau_n)^2, 0 < \tau_1 < \cdots < \tau_N$, a corresponding set (left or right) of positive norming constants

$$\alpha_n^+ := \left(\int_{-\infty}^{\infty} e^2(\rho_n, x)dx \right)^{-1} \quad \text{or} \quad \alpha_n^- := \left(\int_{\infty}^{\infty} g^2(\rho_n, x)dx \right)^{-1}$$

and the reflection coefficient (left or right) $s^+(\rho)$ or $s^-(\rho)$, $\rho \in \mathbb{R}$. The sets

$$J^{\pm} = \left\{ s^{\pm}(\rho), \ \rho \in \mathbb{R}; \ \lambda_n, \alpha_n^{\pm}, \ n = 1, \ldots, N \right\} \tag{6.3}$$

are called the *scattering data*. Thus, the scattering problem consists, given a potential $q(x)$, in finding the associated scattering data J^+ or J^-. Once J^+ is found, it is easy to compute J^- and vice versa [172].

The *inverse scattering problem* is stated as follows. Given scattering data J^+ or J^-, find $q(x)$.

Necessary and sufficient conditions for a set J^+ or J^- to be the scattering data for a potential q can be found in [172, p. 238].

Needless to say, the direct and the inverse scattering problems are computationally challenging. Solution of the direct scattering problem reduces in fact to the computation of the Jost solutions $e(\rho, x)$, $g(\rho, x)$, and the functional series representations for $e(\rho, x)$ (Chap. 10) and $g(\rho, x)$ (Chap. 16) lead to a simple and practical method for solving the direct scattering problem. The numerical implementation of this approach is explained in Chap. 12.

A simple and direct method for the solution of the inverse scattering problem is presented in Chap. 16. It was introduced in [104] and, similarly to the methods of solution of other inverse problems considered in Chaps. 13–15, it reduces the inverse scattering problem on the line to a system of linear algebraic equations. The only difference is that the system is obtained with the aid of the Gelfand-Levitan-Marchenko equation, but remarkably, the whole scheme using a functional series representation for the corresponding integral transmutation kernel and the fact that it suffices to know the first coefficient of the series for recovering the potential q remains in force in this case as well.

The direct and the inverse scattering problems on the line have become especially important because of their usefulness for solving certain nonlinear partial differential equations of mathematical physics in one space and one time dimension. We briefly explain this application of the direct and inverse scattering problems in the next chapter.

Inverse Scattering Transform Method

7

In this chapter we give a brief description of a spectacular and very important application of the direct and inverse scattering problems formulated in the previous chapter. Here the following remark is pertinent. Among specialists in applied fields there can be encountered an opinion that the mathematical theory of the inverse spectral problems has a limited applicability because in practice only a very restricted part of the spectral data can be available from measurements, typically this concerns inverse Sturm-Liouville problems on a finite interval. Needless to say that the powerful mathematical theory specifying conditions for the existence and uniqueness of the solution as well as offering certain methods for its computation is of great value also in practical problems. However, the inverse scattering transform method gives us a beautiful example of application of the direct and inverse spectral problems for which "all the data" are indeed required and can be known in practice. Here the quotes are used of course because when it comes to computation finite instead of infinite intervals can be considered and hence the knowledge of the reflection coefficient on a finite interval only can be used for recovering the potential. Nevertheless, up to this refinement, the complete spectral data can be supposed to be known.

The inverse scattering transform method for solving nonlinear partial differential equations was discovered in 1967 in the paper [73] in application to the *Korteweg-de Vries equation*

$$u_t + u_{xxx} + 6uu_x = 0,$$

which models shallow-water waves and admits solitary wave solutions. The method was successfully applied in 1972 in [173] to another important equation of mathematical physics, the *nonlinear Schrödinger equation*, whose principal applications are to the

V. V. Kravchenko, *Direct and Inverse Sturm-Liouville Problems*,
Frontiers in Mathematics, https://doi.org/10.1007/978-3-030-47849-0_7

propagation of light in nonlinear optical fibers (see, e.g., [157]):

$$i u_t + u_{xx} \pm 2ku \, |u|^2 = 0.$$

While solving the Korteweg-de Vries equation by the inverse scattering transform method reduces precisely to solving the direct and inverse scattering problems from the previous chapter, in the case of the nonlinear Schrödinger equation the appropriate scattering problems correspond to the so-called Zakharov-Shabat system instead of Eq. (6.1) although this difference is not essential.

After those first works, the inverse scattering transform method found many other applications related to a large number of nonlinear evolution partial differential equations of mathematical physics. We refer to the books [1–3] covering parts of this wide topic.

Schematically, for the Korteweg-de Vries equation the inverse scattering transform method can be presented as follows. Given $q(x)$. It is required to solve the Korteweg-de Vries equation with the initial condition $u(x, 0) = q(x)$. The first step is to solve the direct scattering problem for $q(x)$. This produces the scattering data corresponding to $t = 0$ (see the first line of Diagram 1). Then a simple linear evolution law is applied to the scattering data which gives us the scattering data for all $t > 0$. Finally, for every instant $t > 0$ an inverse scattering problem should be solved for a corresponding set of the scattering data $J^+(t)$ or $J^-(t)$. The solution of this inverse problem is precisely the sought-for solution $u(x, t)$.

Diagram 1   Schematic representation of the IST

$$u(x, 0) = q(x) \quad \underset{\text{direct scattering problem}}{\xrightarrow{-y'' + q(x)y = \lambda y}} \quad J^+ \text{ or } J^-(t = 0)$$

$$\Big\downarrow \text{ linear evolution law}$$

$$u(x, t) \quad \underset{\xleftarrow{\text{inverse scattering problem}}} \quad J^+ \text{ or } J^-(t)$$

Often this procedure is referred to as direct and inverse a *nonlinear Fourier transform* [4]. Let us clarify the analogy with the usual linear Fourier transform method considering a linear evolution partial differential equation

$$u_t = -i A (-i \partial_x) u, \quad t > 0, \tag{7.1}$$

where $A (-i \partial_x) = a_0 + a_1 (-i \partial_x) + \cdots + a_n (-i \partial_x)^n$ is a linear differential expression with constant coefficients. We are interested in solving the Cauchy problem for (7.1) with the initial condition $u(x, 0) = q(x)$.

Application of the Fourier transform leads to a problem for the image of u,

$$\widehat{u}(\kappa, t) = \int_{-\infty}^{\infty} u(x, t) e^{-i\kappa x} dx,$$

governed by the differential equation

$$\widehat{u}_t(\kappa, t) + i A(\kappa) \widehat{u}(\kappa, t) = 0.$$

Its solution has the form

$$\widehat{u}(\kappa, t) = \widehat{u}(\kappa, 0) e^{-i A(\kappa) t},$$

where

$$\widehat{u}(\kappa, 0) = \int_{-\infty}^{\infty} u(x, 0) e^{-i\kappa x} dx = \widehat{q}(\kappa).$$

Thus, the sought-for solution $u(x, t)$ of (7.1) is obtained by applying the inverse Fourier transform:

$$u(x, t) = \frac{1}{2\pi} \int_{-\infty}^{\infty} e^{i\kappa x} \widehat{q}(\kappa) e^{-i A(\kappa) t} d\kappa.$$

The Fourier transform method for solving the Cauchy problem is schematically presented in Diagram 2.

Diagram 2 Schematic representation of the Fourier transform method

$$u(x, 0) = q(x) \quad \xrightarrow{\text{forward Fourier transform}} \quad \widehat{u}(\kappa, 0) = \widehat{q}(\kappa)$$

$$\Big| \text{ linear evolution law}$$

$$\downarrow$$

$$u(x, t) \quad \xleftarrow{\text{inverse Fourier transform}} \quad \widehat{u}(\kappa, t) = \widehat{u}(\kappa, 0) e^{-i A(\kappa) t}$$

Thus, the inverse scattering transform method is clearly an extension of the Fourier transform method to certain types of nonlinear evolution partial differential equations.

Part II

Transmutation Operators and Series Representations for Solutions of Sturm-Liouville Equations

In this part of the book the main transmutation operators are introduced and functional series representations for their integral kernels are derived. They are used for obtaining functional series representations for solutions of Sturm-Liouville equations satisfying initial conditions at the origin, as well as for Jost solutions of Schrödinger equations with short-range potentials.

Main Transmutation Operators

8

8.1 Transmutation on a Symmetric Interval

Let us consider a solution u of the equation

$$-u'' + q(x)u = \lambda u, \quad -b < x < b \tag{8.1}$$

satisfying the initial conditions

$$u(\rho, 0) = 1, \quad u'(\rho, 0) = i\rho, \tag{8.2}$$

$\rho = \sqrt{\lambda} \in \mathbb{C}$. Here the underlying interval is supposed to be symmetric and q is a complex-valued function belonging to $L_2(-b, b)$. The following important result is well known (see, e.g., [127, 133, 172]).

Theorem 8.1 *There exists a function $K(x, t)$, defined and continuous in the domain $0 \leq |t| \leq |x| \leq b$, such that for all $\rho \in \mathbb{C}$ the solution $u(\rho, x)$ admits the representation*

$$u(\rho, x) = e^{i\rho x} + \int_{-x}^{x} K(x, t)e^{i\rho t} dt \tag{8.3}$$

and

$$K(x, x) = \frac{1}{2} \int_{0}^{x} q(t)dt, \quad K(x, -x) = 0, \quad -b \leq x \leq b. \tag{8.4}$$

V. V. Kravchenko, *Direct and Inverse Sturm-Liouville Problems*,
Frontiers in Mathematics, https://doi.org/10.1007/978-3-030-47849-0_8

Of crucial importance is the fact that the integral kernel $K(x, t)$ in (8.3) is independent of ρ. According to formula (8.3), $u(\rho, x)$ can be regarded as a result of application of a Volterra integral operator of second kind to the function $e^{i\rho x}$, which is nothing but a solution of the elementary equation $-y'' = \lambda y$ with the same initial conditions (8.2). We denote this Volterra integral operator by T:

$$Tv(x) := v(x) + \int_{-x}^{x} K(x, t)v(t)dt.$$

When $q \in C^1[-b, b]$ the function $K(x, t)$ satisfies the partial differential equation of hyperbolic type

$$\left(\frac{\partial^2}{\partial x^2} - q(x)\right) K(x, t) = \frac{\partial^2}{\partial t^2} K(x, t) \tag{8.5}$$

in the domain $0 \le |t| < |x| \le b$ which combined with the conditions (8.4) represents a *Goursat problem* satisfied by $K(x, t)$.

For every $v \in C^2[-b, b]$,

$$\left(\frac{d^2}{dx^2} - q(x)\right) Tv = T \frac{d^2}{dx^2} v,$$

and thus T not only transforms solutions of the elementary equation into solutions of (8.1), but also the elementary operator itself into the Sturm-Liouville operator $\frac{d^2}{dx^2} - q(x)$.

Example 8.1 Let $q \equiv -c$, where c is a positive constant. Then

$$K(x, t) = -\frac{1}{2} \frac{\sqrt{c(x^2 - t^2)} J_1\left(\sqrt{c(x^2 - t^2)}\right)}{x - t}.$$

Since the operator T is a Volterra integral operator of the second kind with a continuous kernel, it possesses an inverse operator T^{-1} of the same form, but with a different kernel. Hence,

$$\left(\frac{d^2}{dx^2} - q(x)\right) v = T \frac{d^2}{dx^2} T^{-1} v$$

for all $v \in C^2[-b, b]$, which shows that the Sturm-Liouville operator $\frac{d^2}{dx^2} - q(x)$ is similar to the second derivative operator $\frac{d^2}{dx^2}$, and the similarity transformation is performed by the operators T and T^{-1}. Such operators transforming one linear differential operator into another, bounded, invertible and with bounded inverse, will be called *transmutation operators*, or simply *transmutations*. Very often in the mathematical literature transmutation operators are called *transformation operators*. Here we keep to the original term coined by J. Delsarte and J. L. Lions [62].

The concept of a transmutation operator appeared first in the work of J. Delsarte [60, 61]. In [140] A. Ya. Povzner proved that the transmutation operator of the type under consideration can be realized in the form of a Volterra integral operator of second kind. It is precisely that result made transmutation operators especially valuable in the spectral theory of second-order linear differential operators [127, 128, 133].

Sometimes it is more convenient to consider a slightly modified transmutation operator on a symmetric interval. Let f be a solution of the equation

$$f'' - q(x)f = 0, \quad -b < x < b,$$

such that

$$f(0) = 1.$$

Denote $h := f'(0)$. We emphasize that the function f can be complex-valued and thus $h \in \mathbb{C}$.

Theorem 8.2 *There exists a function $K_h(x, t)$, defined and continuous in the domain $0 \le |t| \le |x| \le b$, such that the operator $\mathbf{T}_h$ defined on $C[-b, b]$ by the formula*

$$\mathbf{T}_h v(x) = v(x) + \int_{-x}^{x} K_h(x, t)v(t)dt$$

maps each solution v of the equation $v'' + \rho^2 v = 0$ into a solution u of Eq. (8.1), with the following correspondence of the initial values: $u(0) = v(0)$, $u'(0) = v'(0) + hv(0)$, for all $\rho \in \mathbb{C}$. In particular,

$$\mathbf{T}_h[1] = f.$$

The function $K_h(x, t)$ satisfies the following conditions

$$K_h(x, x) = \frac{h}{2} + \frac{1}{2}\int_0^x q(s)\,ds, \qquad K_h(x, -x) = \frac{h}{2}, \qquad 0 \le x \le b. \tag{8.6}$$

For the proof of this fact we refer to [132, Theorem 3.1.1]. A slightly different proof and references to earlier publications are given in [115]. Obviously, $K_0(x, t) = K(x, t)$.

8.2   Transmutations with Boundary Conditions at the Origin

It is often convenient to use transmutation operators with boundary conditions at the origin.
For example, the solution $\varphi(\rho, x)$ of the Cauchy problem (3.3), (3.4) can be obtained from
the function $\cos \rho x$ (which is a solution of the Cauchy problem $-y'' = \lambda y$, $y(0) = 1$,
$y'(0) = 0$) by acting on it with a transmutation operator T_c given by

$$\varphi(\rho, x) = T_c[\cos \rho x] := \cos \rho x + \int_0^x G(x, t) \cos \rho t \, dt.$$

The equality

$$\left(\frac{d^2}{dx^2} - q(x)\right) T_c v = T_c \frac{d^2}{dx^2} v$$

holds for all the functions $v \in C^2[0, b]$ that satisfy the condition $v'(0) = 0$. The kernel
$G(x, t)$ satisfies Eq. (8.5) in the domain $0 \le t < x \le b$ as well as the conditions

$$G(x, x) = h + \frac{1}{2} \int_0^x q(t) \, dt, \quad \left.\frac{\partial}{\partial t} G(x, t)\right|_{t=0} = 0. \tag{8.7}$$

Similarly, the solution $\psi(\rho, x)$ of Eq. (3.3) with the initial conditions

$$\psi(\rho, 0) = 0, \quad \psi'(\rho, 0) = 1$$

is obtained from the function $\frac{\sin \rho x}{\rho}$ (which is a solution of the Cauchy problem $-y'' = \lambda y$,
$y(0) = 0$, $y'(0) = 1$) with the aid of a transmutation operator T_s given by

$$\psi(\rho, x) = T_s\left[\frac{\sin \rho x}{\rho}\right] := \frac{\sin \rho x}{\rho} + \int_0^x S(x, t) \frac{\sin \rho t}{\rho} \, dt$$

where the kernel $S(x, t)$ satisfies Eq. (8.5) in the domain $0 \le t < x \le b$ as well as the
conditions

$$S(x, x) = \frac{1}{2} \int_0^x q(t) \, dt, \quad S(x, 0) = 0. \tag{8.8}$$

The equality

$$\left(\frac{d^2}{dx^2} - q(x)\right) T_s v = T_s \frac{d^2}{dx^2} v$$

holds for all the functions $v \in C^2[0, b]$ that satisfy the condition $v(0) = 0$.

The relations between the kernels of the transmutation operators introduced above are given by the equalities (see [133])

$$G(x,t) = h + K(x,t) + K(x,-t) + h \int_t^x (K(x,s) - K(x,-s))\, ds,$$

$$G(x,t) = K_h(x,t) + K_h(x,-t) \tag{8.9}$$

and

$$S(x,t) = K_h(x,t) - K_h(x,-t). \tag{8.10}$$

Consequently,

$$K_h(x,t) = \frac{1}{2}(G(x,t) + S(x,t)).$$

Simple relations hold between the kernels corresponding to different values of h.

Theorem 8.3 ([132] (See Also [41] and [113])) *The integral kernels $K_h(x,t)$ and $K_{h_1}(x,t)$ are related by the equality*

$$K_h(x,t) = \frac{h - h_1}{2} + K_{h_1}(x,t) + \frac{h - h_1}{2} \int_t^x \left(K_{h_1}(x,s) - K_{h_1}(x,-s) \right) ds. \tag{8.11}$$

The corresponding operators $\mathbf{T}_h$ and $\mathbf{T}_{h_1}$ are related by the equality

$$\mathbf{T}_h v = \mathbf{T}_{h_1}\left[v(x) + \frac{h - h_1}{2} \int_{-x}^x v(t)\, dt \right], \tag{8.12}$$

which is for any $v \in C[-b,b]$.

In particular,

$$K_h(x,t) = \frac{h}{2} + K(x,t) + \frac{h}{2} \int_t^x \left(K(x,s) - K(x,-s) \right) ds. \tag{8.13}$$

and

$$\mathbf{T}_h v = T\left[v(x) + \frac{h}{2} \int_{-x}^x v(t)\, dt \right]. \tag{8.14}$$

8.3   The Transmutation Operators with Conditions at Infinity

The Jost solution $e(\rho, x)$ introduced in Chap. 4 (under the condition (4.2)) admits the representation

$$e(\rho, x) = e^{i\rho x} + \int_x^\infty A(x, t) e^{i\rho t} dt, \tag{8.15}$$

where A is a real-valued function such that

$$A(x, x) = \frac{1}{2} \int_x^\infty q(t) dt \tag{8.16}$$

and $A(x, \cdot) \in L_2(x, \infty)$. Formula (8.15) was obtained in [126] and is sometimes referred to as *Levin's representation of the Jost solution* (see, e.g., [51, Ch. 5, Sect. 1]).

Analogously, the Jost solution $g(\rho, x)$ introduced in Chap. 6 (under the condition (6.2)) admits the representation

$$g(\rho, x) = e^{-i\rho x} + \int_{-\infty}^x B(x, t) e^{-i\rho t} dt \tag{8.17}$$

where B is a real-valued function such that

$$B(x, x) = \frac{1}{2} \int_{-\infty}^x q(t) dt$$

and $B(x, \cdot) \in L_2(-\infty, x)$.

More on the properties of the kernels A and B can be found in [127].

Construction of Transmutations and Series Representations for Solutions

9.1 A Mapping Property of Transmutations

Let f be a solution of the equation $f'' - q(x)f = 0$ on the interval $(0, b)$ and $f(0) = 1$, $f'(0) = h$. Thus, $f(x) = \varphi(0, x)$ which is the solution of the Cauchy problem (3.3) and (3.4) with $\lambda = 0$.

The next theorem [41] shows that the formal powers are images of the powers of the independent variable under the action of an appropriate transmutation operator.

Theorem 9.1 *Let q be a complex-valued function of the independent real variable $x \in [0, b]$, $q \in L_1(0, b)$, and let $f(x) = \varphi(0, x)$ be a solution of the Cauchy problem (3.3) and (3.4) with $\lambda = 0$. Let φ_k, $k \in \mathbb{N}_0 := \mathbb{N} \cup \{0\}$ be the formal powers defined by (2.5). Then*

$$\varphi_k(x) = T_c[x^k], \quad \text{if } k \in \mathbb{N}_0 \text{ is even,}$$

$$\varphi_k(x) = T_s[x^k], \quad \text{if } k \in \mathbb{N} \text{ is odd,}$$

and

$$\varphi_k(x) = \mathbf{T}_h\left[x^k\right] \qquad \text{for any } k \in \mathbb{N} \cup \{0\}. \tag{9.1}$$

Proof Let us prove (9.1). The other two equalities are consequence of (9.1).

Consider the solution $u_h(\rho, x)$ of the equation $-u'' + q(x)u = \lambda u$, $0 < x < b$ (or $-b < x < b$) satisfying the initial conditions $u_h(\rho, 0) = 1$, $u'_h(\rho, 0) = \rho + h$, where the

V. V. Kravchenko, *Direct and Inverse Sturm-Liouville Problems*, Frontiers in Mathematics, https://doi.org/10.1007/978-3-030-47849-0_9

constant h is from the condition (3.4) and hence $h = f'(0)$. By Theorem 8.2,

$$u_h(\rho, x) = e^{\rho x} + \int_{-x}^{x} K_h(x, t)e^{\rho t}\, dt. \tag{9.2}$$

In terms of the solutions y_1 and y_2 from Theorem 2.1 with $a = x_0 = 0$,

$$u_h(\rho, x) = y_1(x) + \rho y_2(x),$$

and hence

$$u_h(\rho, x) = \sum_{k=0}^{\infty} \frac{\rho^k}{k!} \varphi_k(x).$$

Substituting this SPPS representation of $u_h(\rho, x)$ into (9.2) together with the Maclaurin series for $e^{\rho x}$, we obtain the equality

$$\sum_{k=0}^{\infty} \frac{\rho^k}{k!} \varphi_k(x) = \sum_{k=0}^{\infty} \frac{\rho^k}{k!} \mathbf{T}_h \left[x^k \right],$$

which in view of the uniform convergence of both series on any compact subset of the complex ρ-plane, yields (9.1).   ∎

The importance of this mapping property is explained by the following observation. Even when the integral transmutation kernel is not known, it is always known what are the images of the integer nonnegative powers of the independent variable under the action of the transmutation operator. Namely, they are the formal powers $\varphi_k(x)$, for the construction, of which a simple recurrent integration procedure is available (see Sect. 2.2). This opens the way to the use of the transmutation operator even when it is not known.

For example, with the aid of the transmutation operator *complete systems of solutions* of the *Laplace equation*, the *heat equation* and the *wave equation* can be transmuted into corresponding complete systems of solutions of the equations

$$\left(\frac{\partial^2}{\partial y^2} + \frac{\partial^2}{\partial x^2} - q(x) \right) u(x, y) = 0, \tag{9.3}$$

$$\left(\frac{\partial}{\partial t} + \frac{\partial^2}{\partial x^2} - q(x) \right) u(x, t) = 0 \tag{9.4}$$

and

$$\left(-\frac{\partial^2}{\partial t^2} + \frac{\partial^2}{\partial x^2} - q(x) \right) u(x, t) = 0, \tag{9.5}$$

respectively. This idea of obtaining in such way complete systems of solutions of partial differential equations with variable coefficients was explored, e.g., in [56] and [28] where for this purpose it was suggested to compute the integral transmutation kernel by the method of successive approximations. This approach did not lead to a successful numerical technique for solving corresponding boundary value problems for the partial differential equations, simply because it was computationally expensive and did not guarantee accurate results.

The mapping property proved here allows us to transmute, for example, such complete systems of solutions of the three basic partial differential equations as the *harmonic polynomials*, the *heat polynomials* and the *wave polynomials* into corresponding systems of solutions of (9.3)–(9.5) with minimal computational effort (concerning the completeness of the wave polynomials see [115] and [116]). This idea was developed in [39–41, 107], and [98] and thus the methods of solution of boundary value problems that had been available for partial differential equations with constant coefficients only (because they required the knowledge of complete systems of solutions of the equation) became applicable to equations with variable coefficients of the type (9.3)–(9.5).

The results of the next section allow one, among other applications, to transmute instead of polynomial-type systems of functions, the systems of exponential functions and thus open the possibility to transmute Fourier series into corresponding solutions of partial differential equations with variable coefficients. Besides the applications of these results which are presented in subsequent chapters, we mention [99] and [91], where the transmuted systems of exponential functions were used to solve two different applied problems.

9.2   Fourier–Legendre Series for $K_h(x, t)$ and NSBF for Solutions

In this section we follow [106] and derive a Fourier–Legendre series representation for the kernel $K_h(x, t)$ and, as a corollary, a representation for the solution of the Sturm–Liouville equation in terms of a Neumann series of Bessel functions (NSBF). According to [170, Chapter XVI], "Any series of the type

$$\sum_{n=0}^{\infty} a_n J_{\nu+n}(z)$$

is called a Neumann series, although in fact Neumann considered only the special type of series for which ν is an integer; the investigation of the more general series is due to Gegenbauer". The papers by Carl Gottfried Neumann and Leopold Bernhard Gegenbauer were published in 1867 and 1877, respectively. Since then the NSBF were studied in numerous publications (see [170, 171] and the recent publications [17, 18] on the subject, and references therein).

An interesting research reported in [53] and [67] contains a representation of solutions of Sturm–Liouville equations in the form of NSBF different from the representation presented below. The series from [53] and [67] does not enjoy the uniformity with respect to ρ, in contrast to our representation, and the convergence of the series which is guaranteed on a certain interval of x for holomorphic q only is achieved due to the exponential decay of the values of the spherical Bessel functions $j_n(z)$ when $n \to \infty$. Apart from that previous work, to our best knowledge, the NSBF have not been used to represent solutions of a general linear differential equation. In view of the attractive features of the representation presented here, the Neumann series of Bessel functions, should be considered as a natural and important object of study in the theory of linear differential equations.

Note that the definition of the transmutation operator $\mathbf{T}_h$ requires that the integral kernel K_h be known only in the regions $R_1 := \{0 \le x \le b, \ |t| \le x\}$ and $R_2 := \{-b \le x \le 0, \ |t| \le |x|\}$. Moreover, these two regions are independent in the following sense. The integral kernel K_h in R_1 depends only on the values of the potential q on $[0, b]$ and does not depend on values for $x < 0$, and in R_2 it depends on the values on $[-b, 0]$ and does not depend on the values for $x > 0$. Obtaining the value of $\mathbf{T}_h[v](x)$ for $x \ge 0$ does not require that K_h be known on R_2. The same happens with the formal powers $\{\varphi_k\}$ and $\{\psi_k\}$, whose values on $[0, b]$ are independent of the values of the potential q on $[-b, 0)$. Therefore, from now on we restrict the presentation to the segment $[0, b]$; all the results for the segment $[-b, 0]$ are similar.

Let P_n denote the *Legendre polynomial* of order n and let $l_{k,n}$ be the corresponding coefficient of x^k, that is $P_n(x) = \sum_{k=0}^{n} l_{k,n} x^k$.

Theorem 9.2 ([106]) *Let $q \in L_2(0, b)$. The transmutation kernel $K_h(x, t)$ from Theorem 8.2 has the form*

$$K_h(x, t) = \sum_{j=0}^{\infty} \frac{\beta_j(x)}{x} P_j \left(\frac{t}{x} \right), \tag{9.6}$$

where for each fixed $x \in (0, b]$ the series converges in the norm of $L_2(-x, x)$ and uniformly with respect to $t \in [-x, x]$ when $q \in C[0, b]$. The coefficients β_j have the form

$$\beta_j(x) = \frac{2j+1}{2} \left(\sum_{k=0}^{j} \frac{l_{k,j} \varphi_k(x)}{x^k} - 1 \right). \tag{9.7}$$

Proof Since the kernel $K_h \in C([-b, b] \times [-b, b])$, for any $x \in (0, b]$ it admits (see, e.g., [161]) a Fourier–Legendre series expansion of the form $\sum_{j=0}^{\infty} A_j(x) P_j \left(\frac{t}{x} \right)$ that converges in the norm of $L_2(-x, x)$ where for convenience we consider $A_j(x) = \frac{\alpha_j(x)}{x}$ (when $q \in$

$C[0, b]$ the function K_h is continuously differentiable with respect to both arguments and hence its Fourier–Legendre series converges uniformly, see [106] for details).

Multiplying (9.6) by $P_n\left(\frac{t}{x}\right)$ and integrating, we obtain

$$\int_{-x}^{x} K_h(x, t) P_n\left(\frac{t}{x}\right) dt = \sum_{j=0}^{\infty} \frac{\beta_j(x)}{x} \int_{-x}^{x} P_j\left(\frac{t}{x}\right) P_n\left(\frac{t}{x}\right) dt = \frac{2}{2n+1} \beta_n(x).$$

(9.8)

Hence,

$$\beta_n(x) = \frac{2n+1}{2} \int_{-x}^{x} K_h(x, t) P_n\left(\frac{t}{x}\right) dt = \frac{2n+1}{2} \sum_{k=0}^{n} \int_{-x}^{x} K_h(x, t) l_{k,n} \left(\frac{t}{x}\right)^k dt$$

$$= \frac{2n+1}{2} \sum_{k=0}^{n} \frac{l_{k,n}}{x^k} \int_{-x}^{x} K_h(x, t) t^k dt = \frac{2n+1}{2} \sum_{k=0}^{n} \frac{l_{k,n}}{x^k} \left(\mathbf{T}_h\left[x^k\right] - x^k\right).$$

Using Theorem 9.1, we obtain

$$\beta_n(x) = \frac{2n+1}{2} \sum_{k=0}^{n} \frac{l_{k,n}}{x^k} \left(\varphi_k(x) - x^k\right),$$

which yields (9.7) because $P_n(1) = 1$.   ∎

An immediate corollary of this theorem is a Fourier–Legendre series representation for the transmutation kernels $G(x, t)$ and $S(x, t)$ introduced in Sect. 8.2.

Corollary 9.1 *The integral transmutation kernels $G(x, t)$ and $S(x, t)$ admit the following series representations:*

$$G(x, t) = \sum_{n=0}^{\infty} \frac{g_n(x)}{x} P_{2n}\left(\frac{t}{x}\right), \quad 0 < t \le x \le b,$$

and

$$S(x, t) = \sum_{n=0}^{\infty} \frac{s_n(x)}{x} P_{2n+1}\left(\frac{t}{x}\right), \quad 0 < t \le x \le b,$$

where $g_n = 2\beta_{2n}$ and $s_n = 2\beta_{2n+1}$. For each fixed $x \in (0, b]$, the series converge in the norm of $L_2(0, x)$.

Proof The proof follows immediately from Theorem 9.2, formulas (8.9), (8.10) and the following property of the Legendre polynomials: $P_n(-x) = (-1)^n P_n(x)$. ∎

We give without proof the following useful proposition that establishes an estimate for the remainder of the series (9.6). Denote the partial sum of the series (9.6) by

$$K_{h,N}(x, t) := \sum_{n=0}^{N} \frac{\beta_n(x)}{x} P_n\left(\frac{t}{x}\right).$$

Proposition 9.1 ([106]) *Suppose that $q \in C^p[0, b]$ and define*

$$M := \max_{0 \leq x \leq b, \, |t| \leq x} \left| \partial_t^{p+1} K_h(x, t) \right|. \tag{9.9}$$

Then for all $N > p$, $0 < x \leq b$, and $|t| \leq x$

$$\left| K_h(x, t) - K_{h,N}(x, t) \right| \leq \frac{c_p M x^{p+1}}{N^{p+1/2}}, \tag{9.10}$$

where the constant c_p does not depend on q and N.

The next theorem we (following [106]) formulate and prove in the case $q \in C[0, b]$; in Remark 9.1 below we discuss the case $q \in L_2(0, b)$.

Theorem 9.3 ([106]) *Let $q \in C[0, b]$. The solutions $\varphi(\rho, x)$ and $s(\rho, x)$ of the equation*

$$-y'' + q(x)y = \rho^2 y, \quad \rho \in \mathbb{C}, \quad 0 < x < b, \tag{9.11}$$

satisfying the initial conditions

$$\varphi(\rho, 0) = 1, \quad \varphi'(\rho, 0) = h$$

and

$$s(\rho, 0) = 0, \quad s'(\rho, 0) = \rho$$

admit the representations

$$\varphi(\rho, x) = \cos \rho x + \sqrt{\frac{2\pi}{\rho x}} \sum_{n=0}^{\infty} (-1)^n \beta_{2n}(x) J_{2n+1/2}(\rho x)$$

$$= \cos \rho x + 2 \sum_{n=0}^{\infty} (-1)^n \beta_{2n}(x) j_{2n}(\rho x) \tag{9.12}$$

and

$$s(\rho, x) = \sin \rho x + \sqrt{\frac{2\pi}{\rho x}} \sum_{n=0}^{\infty} (-1)^n \beta_{2n+1}(x) J_{2n+3/2}(\rho x)$$

$$= \sin \rho x + 2 \sum_{n=0}^{\infty} (-1)^n \beta_{2n+1}(x) j_{2n+1}(\rho x),$$

(9.13)

where j_k stands for the spherical Bessel function of order k defined as $j_k(z) := \sqrt{\frac{\pi}{2z}} J_{k+\frac{1}{2}}(z)$ (and J_ν stands for the Bessel function of order ν). The series converge uniformly with respect to x on $[0, b]$ and converge uniformly with respect to ρ on any compact subset of the complex ρ-plane. Moreover, the functions

$$\varphi_N(\rho, x) = \cos \rho x + 2 \sum_{n=0}^{[N/2]} (-1)^n \beta_{2n}(x) j_{2n}(\rho x)$$

(9.14)

and

$$s_N(\rho, x) = \sin \rho x + 2 \sum_{n=0}^{[(N-1)/2]} (-1)^n \beta_{2n+1}(x) j_{2n+1}(\rho x)$$

(9.15)

obey the estimates

$$\left| \varphi(\rho, x) - \varphi_N(\rho, x) \right| \leq 2|x| \varepsilon_N(x) \quad \text{and} \quad |s(\rho, x) - s_N(\rho, x)| \leq 2|x| \varepsilon_N(x)$$

(9.16)

for any $\rho \in \mathbb{R}$, $\rho \neq 0$, and

$$\left| \varphi(\rho, x) - \varphi_N(\rho, x) \right| \leq \frac{2\varepsilon_N(x) \sinh(Cx)}{C} \quad \text{and} \quad |s(\rho, x) - s_N(\rho, x)| \leq \frac{2\varepsilon_N(x) \sinh(Cx)}{C}$$

(9.17)

for any $\rho \in \mathbb{C}$, $\rho \neq 0$ belonging to the strip $|\text{Im } \rho| \leq C$, $C \geq 0$, where ε_N is a sufficiently small nonnegative function such that $\left| K_h(x, t) - K_{h,N}(x, t) \right| \leq \varepsilon_N(x)$, the existence of which is guaranteed by Theorem 9.2.

Proof According to Theorem 8.2, the solutions $\varphi(\rho, x)$ and $s(\rho, x)$ admit the representations

$$\varphi(\rho, x) = \cos \rho x + \int_{-x}^{x} K_h(x, t) \cos \rho t \, dt$$

(9.18)

and

$$s(\rho, x) = \sin \rho x + \int_{-x}^{x} K_h(x, t) \sin \rho t \, dt. \tag{9.19}$$

Substitution of $K_h(x, t)$ in the form of the series (9.6) leads to the formulas

$$\varphi(\rho, x) = \cos \rho x + \sum_{j=0}^{\infty} \frac{\beta_j(x)}{x} \int_{-x}^{x} P_j \left(\frac{t}{x} \right) \cos \rho t \, dt$$

$$= \cos \rho x + \sum_{j=0}^{\infty} \beta_j(x) \int_{-1}^{1} P_j(y) \cos (\rho x y) \, dy$$

and

$$s(\rho, x) = \sin \rho x + \sum_{j=0}^{\infty} \beta_j(x) \int_{-1}^{1} P_j(y) \sin (\rho x y) \, dy.$$

Using formula 2.17.7 in [141, p. 433],

$$\int_0^a \left\{ \begin{matrix} P_{2n+1} \left(\frac{y}{a} \right) \cdot \sin by \\ P_{2n} \left(\frac{y}{a} \right) \cdot \cos by \end{matrix} \right\} dy = (-1)^n \sqrt{\frac{\pi a}{2b}} J_{2n+\delta+1/2}(ab), \quad \delta = \left\{ \begin{matrix} 1 \\ 0 \end{matrix} \right\}, \quad a > 0,$$

we obtain the representations (9.12) and (9.13).

The convergence of the series with respect to ρ can be established using the fact that for each x the series represent the Neumann series (see, e.g., [170] and [171]). Indeed, $\varphi(\rho, x) - \cos \rho x$ (and similarly, $\rho (s(\rho, x) - \sin \rho x)$), regarded as a function of the complex variable ρ, is entire and, as the radius of convergence of the Neumann series coincides [170, pp. 524–526] with the radius of convergence of its associated power series (obtained from the SPPS representation), we conclude that the series (9.12) and (9.13) converge uniformly on any compact subset of the complex ρ-plane of the variable ρ.

Consider a complex $\rho \neq 0$ belonging to the strip $|\mathrm{Im}\, \rho| \leq C$. Then

$$|\varphi(\rho, x) - \varphi_N(\rho, x)| \leq \int_{-x}^{x} \left| K_h(x, t) - K_{h,N}(x, t) \right| |\cos \rho t| \, dt \leq 2\varepsilon_N(x) \int_0^x |\cos \rho t| \, dt$$

$$\leq \varepsilon_N(x) \int_0^x \left(e^{|\mathrm{Im}\, \rho|t} + e^{-|\mathrm{Im}\, \rho|t} \right) dt = 2\varepsilon_N(x) \int_0^x \cosh \left(|\mathrm{Im}\, \rho| t \right) dt$$

$$= \frac{2\varepsilon_N(x) \, \sinh(|\mathrm{Im}\, \rho| x)}{|\mathrm{Im}\, \rho|}. \tag{9.20}$$

Since the function $\sinh(\xi x)/\xi$ is monotonically increasing in both variables when $\xi, x \geq 0$, we obtain the required first inequality in (9.17). The second inequality in (9.17) and the inequalities (9.16) are proved similarly.

The uniform convergence of the series (9.12) and (9.13) with respect to the variable x follows directly from the inequalities (9.16) and (9.17) and the estimate (9.10), valid at least for $p = 0$.   ∎

Remark 9.1 The representations (9.12) and (9.13) remain valid in the case $q \in L_2(0, b)$. However, the proof of the estimates (9.16) and (9.17) should be adjusted with the aid of the Cauchy–Bunyakovsky–Schwarz inequality. Consider the solution $u_h(\rho, x)$ of (9.11) with the initial conditions

$$u_h(\rho, 0) = 1, \quad u'_h(\rho, 0) = i\rho + h.$$

That is,

$$u_h(\rho, x) = \mathbf{T}_h\left[e^{i\rho x}\right] = e^{i\rho x} + \int_{-x}^{x} K_h(x, t)e^{i\rho t}\, dt.$$

Of course, $\varphi(\rho, x) = \frac{1}{2}(u_h(\rho, x) + u_h(-\rho, x))$ and $s(\rho, x) = \frac{1}{2i}(u_h(\rho, x) - u_h(-\rho, x))$. Consider a complex $\rho \neq 0$ belonging to the strip $|\mathrm{Im}\,\rho| \leq C$. Then

$$\left|u_h(\rho, x) - u_{h,N}(\rho, x)\right| = \left|\int_{-x}^{x}\left(K_h(x, t) - K_{h,N}(x, t)\right)e^{i\rho t}\, dt\right|$$

$$\leq \left\|K_h(x, \cdot) - K_{h,N}(x, \cdot)\right\|_{L_2(-x,x)}\left\|e^{i\rho t}\right\|_{L_2(-x,x)}.$$

Since

$$\left\|e^{i\rho t}\right\|_{L_2(-x,x)} = \sqrt{\int_{-x}^{x}e^{-2\,\mathrm{Im}(\rho)t}\, dt} = \sqrt{\frac{\sinh\left(2\,\mathrm{Im}\left(\rho\right)x\right)}{\mathrm{Im}\left(\rho\right)}},$$

we obtain the inequality

$$\left|u_h(\rho, x) - u_{h,N}(\rho, x)\right| \leq \epsilon_N(x)\sqrt{\frac{\sinh\left(2Cx\right)}{C}},$$

with $\epsilon_N(x) := \left\|K_h(x, \cdot) - K_{h,N}(x, \cdot)\right\|_{L_2(-x,x)}$.

Remark 9.2 The inequalities (9.16) and (9.17) are of particular importance when using representations (9.12) and (9.13) to solve spectral problems for the Eq. (9.11) because they guarantee a uniform approximation of eigendata (see [116, Proposition 7.1]), which is illustrated by numerical experiments in Sect. 11.2.

Remark 9.3 Note that for a fixed z the numbers $j_k(z)$ decrease rapidly as $k \to \infty$, see, e.g., [5, (9.1.62)]. Hence, the rate of convergence of the series (9.12) and (9.13) for any fixed ρ (and for bounded subsets $\Omega \subset \mathbb{C}$) is, in fact, exponential.

The following statement, which provides more detailed estimates for the remainders of the series depending on the regularity of the potential q is given here without proof (which can be found in [106]).

Proposition 9.2 ([106]) *Let $x > 0$ be fixed and let $\rho \in \mathbb{C}$ satisfy $|\rho| \le \rho_0$. Suppose that $q \in C^p[0, b]$ for some $p \in \mathbb{N}_0$. Then for all $N > \max\{\rho_0 x, p\}/2$ the remainders of the series (9.12) and (9.13) satisfy*

$$|\varphi(\rho, x) - \varphi_{2N}(\rho, x)| = |\varphi(\rho, x) - \varphi_{2N+1}(\rho, x)| \le \frac{cx^{p+2}e^{|\operatorname{Im}\rho|x}}{(2N+2)^{p+1/2}} \cdot \frac{1}{(2N+2)!} \cdot \left|\frac{\rho_0 x}{2}\right|^{2N+2}$$

and

$$|s(\rho, x) - s_{2N-1}(\rho, x)| = |s(\rho, x) - s_{2N}(\rho, x)| \le \frac{cx^{p+2}e^{|\operatorname{Im}\rho|x}}{(2N+1)^{p+1/2}} \cdot \frac{1}{(2N+1)!} \cdot \left|\frac{\rho_0 x}{2}\right|^{2N+1},$$

where c is a constant depending on q and p only.

9.3   NSBF for Derivatives of Solutions

To solve Sturm–Liouville problems one needs not only convenient representations for the solutions, but also such representations for their derivatives. To provide them, we return to formulas (9.18) and (9.19). Let us assume that $q \in C[0, b]$. Then the kernel $K_h(x, t)$ is continuously differentiable with respect to both arguments, and we can differentiate (9.18) and (9.19) with respect to x. Using (8.6), we obtain the relations

$$\varphi'(\rho, x) = -\rho \sin \rho x + \int_{-x}^{x} K_{h,1}(x, t) \cos \rho t \, dt + \left(h + \frac{1}{2} \int_0^x q(s) \, ds\right) \cos \rho x \qquad (9.21)$$

and

$$s'(\rho, x) = \rho \cos \rho x + \int_{-x}^{x} K_{h,1}(x, t) \sin \rho t \, dt + \frac{1}{2}\left(\int_0^x q(s) \, ds\right) \sin \rho x. \qquad (9.22)$$

Here $K_{h,1}(x, t)$ denotes the derivative of $K_h(x, t)$ with respect to the first variable.

To obtain a convenient representation for the kernel $K_{h,1}(x, t)$, we can apply a procedure similar to that from the previous section. Let us seek $K_{h,1}(x, t)$ in the form

$$K_{h,1}(x, t) = \sum_{j=0}^{\infty} \frac{\gamma_j(x)}{x} P_j\left(\frac{t}{x}\right). \tag{9.23}$$

Then analogously to (9.8) we have

$$\gamma_n(x) = \frac{2n+1}{2} \int_{-x}^{x} K_{h,1}(x, t) P_n\left(\frac{t}{x}\right) dt. \tag{9.24}$$

Differentiating (9.1) (and using (8.6)) we obtain the relations

$$\int_{-x}^{x} K_{h,1}(x, t) t^k dt = \varphi'_k(x) - k x^{k-1} - \frac{1}{2}\left(\left(1 + (-1)^k\right) h + \int_{0}^{x} q(s)\, ds\right) x^k.$$

Next, using (9.24), we deduce that

$$\gamma_n(x) = \frac{2n+1}{2} \sum_{k=0}^{n} \frac{l_{k,n}}{x^k} \int_{-x}^{x} K_{h,1}(x, t) t^k dt$$

$$-\frac{2n+1}{2}\left(\sum_{k=0}^{n} \frac{l_{k,n} \varphi'_k(x)}{x^k} - \frac{n(n+1)}{2x} - \frac{1}{2}\int_{0}^{x} q(s)\, ds - \frac{h}{2}\left(1 + (-1)^n\right)\right), \tag{9.25}$$

where several elementary properties of Legendre polynomials, such as

$$\sum_{k=0}^{n} l_{k,n} = P_n(1) = 1$$

and

$$\sum_{k=1}^{n} k l_{k,n} = P'_n(1) = \frac{n(n+1)}{2},$$

were employed.

Finally, if f does not have zeros in $[0, b]$ (such f always exists, see Remark 2.6), the derivatives φ'_k can be calculated by the formula

$$\varphi'_k = k \psi_{k-1} + \frac{f'}{f} \varphi_k,$$

which follows directly from the definition of the formal powers (2.5) and (2.7). Otherwise, if f has a zero in $[0, b]$, it is convenient to use the formulas from [114].

In general, since $q \in C[0, b]$, the function $K_{h,1}(x, t)$ is continuous in both variables. However, we cannot guarantee any additional smoothness of $K_{h,1}$ as a function of t. It is known that the Fourier–Legendre series of a continuous function may not converge to the function even pointwise. Nevertheless, the series always converges to the function in the L_2 norm.

Theorem 9.4 ([106]) *Let $q \in C[0, b]$. Then the derivatives of the solutions $\varphi(\rho, x)$ and $s(\rho, x)$ of Eq. (9.11) admit the representations*

$$\varphi'(\rho, x) = -\rho \sin \rho x + \left(h + \frac{1}{2}\int_0^x q(s)\, ds\right)\cos \rho x + 2\sum_{n=0}^{\infty}(-1)^n \gamma_{2n}(x) j_{2n}(\rho x)$$

(9.26)

and

$$s'(\rho, x) = \rho \cos \rho x + \frac{1}{2}\left(\int_0^x q(s)\, ds\right)\sin \rho x + 2\sum_{n=0}^{\infty}(-1)^n \gamma_{2n+1}(x) j_{2n+1}(\rho x),$$

(9.27)

where γ_k are defined by (9.25). The series converge uniformly for any $x \in [0, b]$ and converge uniformly with respect to ρ on any compact subset of the complex ρ-plane.

Proof The proof of these representations follows from (9.21) and (9.22) by substitution of (9.23) and arguing similarly to the proof of Theorem 9.3. The uniform convergence with respect to ρ is proved analogously to Theorem 9.3. For the proof of the uniform convergence of the series with respect to x we refer to [106]. ∎

Remark 9.4 Consider the following approximations of the derivatives of the solutions

$$\overset{\circ}{\varphi}_N(\rho, x) = -\rho \sin \rho x + \left(h + \frac{1}{2}\int_0^x q(s)\, ds\right)\cos \rho x + 2\sum_{n=0}^{[N/2]}(-1)^n \gamma_{2n}(x) j_{2n}(\rho x)$$

(9.28)

and

$$\overset{\circ}{s}_N(\rho, x) = \rho \cos \rho x + \frac{1}{2}\left(\int_0^x q(s)\, ds\right)\sin \rho x + 2\sum_{n=0}^{[(N-1)/2]}(-1)^n \gamma_{2n+1}(x) j_{2n+1}(\rho x).$$

(9.29)

Then for the difference $\varphi'(\rho, x) - \overset{\circ}{\varphi}_N(\rho, x)$, as well $s'(\rho, x) - \overset{\circ}{s}_N(\rho, x)$, we obtain similar estimates as in Theorem 9.3 and Remark 9.1. Their proof is analogous.

9.4   A Recurrent Integration Procedure for Computing β_k and γ_k

Formulas (9.7) and (9.25) are simple and therefore attractive. They reduce the computation of the coefficients β_k and γ_k, $k = 0, 1, \ldots$ to the computation of the SPPS formal powers φ_k and ψ_k, $k = 0, 1, \ldots$, which is easily realizable. However, the direct use of (9.7) and (9.25) encounters an important difficulty. The Legendre polynomial coefficients $l_{k,j}$ grow quite fast and thus a limited number of the coefficients β_k and γ_k can be computed in a machine precision arithmetic (numerical examples can be found in [106]). Thus, if one is interested in computing solutions with a higher accuracy and hence more coefficients β_k and γ_k need to be computed, there are two ways to overcome this difficulty: employ a higher precision arithmetic, or derive other recurrent formulas for β_k and γ_k which would be more convenient for computation.

In this section such an alternative recurrent procedure from [106] for calculating the functions β_k and γ_k is derived. It is based on the substitution of the expressions (9.12) and (9.13) of the solutions $\varphi(\rho, x)$ and $s(\rho, x)$ into Eq. (9.11). This approach allows more functions β_k and γ_k to be evaluated numerically (see [106] for numerical examples).

Let us start with the solution $\varphi(\rho, x)$ (see Theorem 9.3). Analogous formulas for $s(\rho, x)$ are given below. We proceed formally and at the end of this section we justify for the case $q \in C^2[0, b]$ the possibility to differentiate termwise all the series and explain why the final formulas remain valid for the general case. Differentiating the solution $\varphi(\rho, x)$ twice, using the formulas

$$j'_k(z) = -j_{k+1}(z) + \frac{k}{z} j_k(z), \quad k = 0, 1, \ldots$$

for the first derivative and

$$j'_k(z) = j_{k-1}(z) - \frac{k+1}{z} j_k(z), \quad k = 1, 2, \ldots$$

for the second derivative, and substituting into (9.11) leads us to the equality

$$2 \sum_{n=0}^{\infty} (-1)^n \left[j_{2n}(\rho x) \left(\beta''_{2n}(x) + \frac{4n}{x} \beta'_{2n}(x) + \frac{2n(2n-1)}{x^2} \beta_{2n}(x) \right) \right.$$

$$\left. -2\rho j_{2n+1}(\rho x) \left(\beta'_{2n}(x) - \frac{\beta_{2n}(x)}{x} \right) \right] = q(x)(\cos \rho x + 2 \sum_{n=0}^{\infty} (-1)^n \beta_{2n}(x) j_{2n}(\rho x)).$$

Combining the terms containing $j_0(\rho x)$ and using the fact that $q(x) = 2(\beta_0''(x) - q(x)\beta_0(x))$, we obtain

$$\left(\cos \rho x - j_0(\rho x)\right)\left(\beta_0''(x) - q(x)\beta_0(x)\right) = -2\rho \sum_{n=0}^{\infty}(-1)^n j_{2n+1}(\rho x)\left(\beta_{2n}'(x) - \frac{1}{x}\beta_{2n}(x)\right)$$

$$+ \sum_{n=1}^{\infty}(-1)^n j_{2n}(\rho x)\left(\beta_{2n}''(x) + \frac{4n}{x}\beta_{2n}'(x) + \frac{2n(2n-1)}{x^2}\beta_{2n}(x) - q(x)\beta_{2n}(x)\right).$$

$$(9.30)$$

The second series can be expressed in terms of odd-index spherical Bessel functions using the equality

$$j_{2n}(\rho x) = \frac{\rho x}{4n+1}\left(j_{2n-1}(\rho x) + j_{2n+1}(\rho x)\right). \tag{9.31}$$

Note additionally that

$$\cos \rho x - j_0(\rho x) = -\rho x \cdot j_1(\rho x). \tag{9.32}$$

Applying (9.32) and (9.31) to (9.30) and dividing by ρx we see that (9.30) can be written as

$$\sum_{n=1}^{\infty} \alpha_n(x) j_{2n-1}(\rho x) = 0, \tag{9.33}$$

where

$$\alpha_n(x) = (-1)^n \left[\frac{1}{4n+1}\left(\beta_{2n}''(x) + \frac{4n}{x}\beta_{2n}'(x) + \left(\frac{2n(2n-1)}{x^2} - q(x)\right)\beta_{2n}(x)\right)\right.$$

$$- \frac{1}{4n-3}\left(\beta_{2(n-1)}''(x) + \frac{4(n-1)}{x}\beta_{2(n-1)}'(x)\right.$$

$$+ \left(\frac{2(n-1)(2(n-1)-1)}{x^2} - q(x)\right)\beta_{2(n-1)}(x)\Big)$$

$$+ 2\left.\left(\frac{1}{x}\beta_{2(n-1)}'(x) - \frac{1}{x^2}\beta_{2(n-1)}(x)\right)\right].$$

Multiplying the equality (9.33) by $j_{2m-1}(\rho x)$, $m = 1, 2, \ldots$, integrating with respect to ρ from 0 to ∞, and using the fact that

$$\int_0^\infty j_{\nu+2n}(y) j_{\nu+2m}(y) \, dy = 0 \tag{9.34}$$

for $n, m \in \mathbb{Z}$ with $n \neq m$ and $m + n + \nu > -1/2$ (c.f., [5, Formula 11.4.6]) we obtain that all coefficients α_n are identically equal to zero.

In order to simplify the equations $\alpha_n(x) = 0$, $n = 1, 2, \ldots$, consider the functions

$$\sigma_{2n}(x) := x^{2n} \beta_{2n}(x), \quad n = 0, 1, \ldots.$$

Then the equations $\alpha_n(x) = 0$ take the form

$$\sigma''_{2n}(x) - q(x)\sigma_{2n}(x) = \frac{4n+1}{4n-3} x^2 \left(\sigma''_{2(n-1)}(x) - q(x)\sigma_{2(n-1)}(x) \right)$$

$$\tag{9.35}$$

$$- 2(4n+1) x \left(\sigma'_{2(n-1)}(x) - \frac{2n-1}{x} \sigma_{2(n-1)}(x) \right).$$

Equations similar to (9.35) can be derived also for the odd coefficients. A calculation similar to that for $\varphi(\rho, x)$ leads to the equality

$$\frac{q(x)}{2} \frac{\sin \rho x}{\rho x} =$$

$$\sum_{n=0}^\infty (-1)^n \left[\frac{j_{2n}(\rho x)}{4n+3} \left(\beta''_{2n+1}(x) + \frac{2(2n+1)}{x} \beta'_{2n+1}(x) + \left(\frac{2n(2n+1)}{x^2} - q(x) \right) \beta_{2n+1}(x) \right) \right.$$

$$+ \frac{j_{2n+2}(\rho x)}{4n+3} \left(\beta''_{2n+1}(x) + \frac{2(2n+1)}{x} \beta'_{2n+1}(x) + \left(\frac{2n(2n+1)}{x^2} - q(x) \right) \beta_{2n+1}(x) \right)$$

$$\left. - 2 j_{2n+2}(\rho x) \left(\frac{1}{x} \beta'_{2n+1}(x) - \frac{1}{x^2} \beta_{2n+1}(x) \right) \right].$$

Noting that $\sin(\rho x)/\rho x = j_0(\rho x)$, we see that the last equality is of the form

$$\sum_{n=0}^\infty \alpha_n(x) j_{2n}(\rho x) = 0,$$

where

$$\alpha_n(x) =$$

$$(-1)^n \left[\frac{1}{4n+3} \left(\beta''_{2n+1}(x) + \frac{2(2n+1)}{x} \beta'_{2n+1}(x) + \left(\frac{2n(2n+1)}{x^2} - q(x) \right) \beta_{2n+1}(x) \right) \right.$$

$$- \frac{1}{4n-1} \left(\beta''_{2n-1}(x) + \frac{2(2n-1)}{x} \beta'_{2n-1}(x) + \left(\frac{(2n-2)(2n-1)}{x^2} - q(x) \right) \beta_{2n-1}(x) \right)$$

$$\left. + 2 \left(\frac{1}{x} \beta'_{2n-1}(x) - \frac{1}{x^2} \beta_{2n-1}(x) \right) \right]$$

and we have taken $\beta_{-1} := 1/2$ to simplify notations for $n = 0$. Using (9.34) one obtains the relations $\alpha_n \equiv 0$ for $n = 0, 1, 2, \ldots$. Introducing $\sigma_{2n+1} = x^{2n+1} \beta_{2n+1}(x)$ we rewrite them in the form

$$\frac{1}{4n+3} \left(\sigma''_{2n+1}(x) - q(x)\sigma_{2n+1}(x) \right)$$

$$= \frac{x^2}{4n-1} \left(\sigma''_{2n-1}(x) - q(x)\sigma_{2n-1}(x) \right) - 2x(\sigma'_{2n-1}(x) - \frac{2n}{x}\sigma_{2n-1}(x)).$$

Combining the even cases with the odd ones we obtain the following sequence of equations for determining the coefficients $\beta_n(x) = x^{-n}\sigma_n(x)$ in the representations of solutions:

$$\frac{1}{2n+1} \left(\sigma''_n(x) - q(x)\sigma_n(x) \right) \tag{9.36}$$

$$= \frac{x^2}{2n-3} \left(\sigma''_{n-2}(x) - q(x)\sigma_{n-2}(x) \right) - 2x(\sigma'_{n-2}(x) - \frac{n-1}{x}\sigma_{n-2}(x)). \tag{9.37}$$

To obtain equations for the coefficients γ_k one has to compare (9.26) and (9.27) with the derivatives of (9.12) and (9.13) and proceed in much the same way as above. We obtain the relations

$$\gamma_0(x) = \beta'_0(x) - \frac{h}{2} - \frac{1}{4} \int_0^x q(s) \, ds,$$

$$\gamma_1(x) = \frac{1}{x} \beta_1(x) + \beta'_1(x) - \frac{3}{4} \int_0^x q(s) \, ds,$$

$$\gamma_n(x) = \frac{n}{x}\beta_n(x) + \beta_n'(x)$$

$$+ \frac{2n+1}{2n-3}\left(\gamma_{n-2}(x) - \beta_{n-2}'(x) + \frac{n-1}{x}\beta_{n-2}(x)\right),\ n = 2, 3, \ldots \qquad (9.38)$$

Note that the last formula holds for $n = 1$ as well if we put $\gamma_{-1} := \frac{1}{4}\int_0^x q(s)\,ds$. Denoting $\tau_n(x) := x^n\gamma_n(x)$ we can rewrite Eq. (9.38) in terms of the functions σ_n:

$$\tau_n(x) = \sigma_n'(x) + \frac{2n+1}{2n-3}x^2\left(\tau_{n-2}(x) - \sigma_{n-2}'(x)\right) + (2n+1)x\sigma_{n-2}(x). \qquad (9.39)$$

Hence the construction of the functions β_n and γ_n for $n = 1, 2, \ldots$ reduces to the solution of a recurrent sequence of inhomogeneous Schrödinger equations (9.36) having the form

$$\sigma_n''(x) - q(x)\sigma_n(x) = h_n(x) \qquad (9.40)$$

with the initial conditions $\sigma_n(0) = \sigma_n'(0) = 0$.

Thus, the following statement is proved.

Proposition 9.3 *The functions $\sigma_n(x) := x^n\beta_n(x)$, where β_n are the coefficients from (9.12) and (9.13), satisfy the recurrent sequence of differential equations (9.36) for $n = 1, 2, \ldots$ with the initial conditions $\sigma_n(0) = \sigma_n'(0) = 0$ and with the first functions given by $\beta_{-1} := 1/2$ and $\beta_0 = (f-1)/2$. The functions $\tau_n(x) := x^n\gamma_n(x)$, where γ_n are the coefficients from (9.26) and (9.27), are given by the sequence of recurrent relations (9.39), with the first functions given by $\gamma_{-1} := \frac{1}{4}\int_0^x q(s)\,ds$ and $\gamma_0 = \frac{f'-h}{2} - \frac{1}{4}\int_0^x q(s)\,ds$.*

Remark 9.5 The values $\sigma_1(x) = \frac{3}{2}(\varphi_1(x) - x)$ and $\tau_1(x) = \frac{3}{2}(\frac{f'\varphi_1+1}{f} - 1 - \frac{x}{2}\int_0^x q(s)\,ds)$ can also be used as the initial values for Proposition 9.3.

Remark 9.6 Let $L := \frac{d^2}{dx^2} - q(x)$. Equations (9.36) and (9.38) can be written in the following somewhat more symmetric form:

$$\frac{1}{x^n}L\left[x^n\beta_n(x)\right] = \frac{2n+1}{2n-3}x^{n-1}L\left[\frac{\beta_{n-2}(x)}{x^{n-1}}\right],$$

$$\gamma_n(x) - \frac{1}{x^n}\left(x^n\beta_n(x)\right)' = \frac{2n+1}{2n-3}\left[\gamma_{n-2}(x) - x^{n-1}\left(\frac{\beta_{n-2}(x)}{x^{n-1}}\right)'\right].$$

The solution of the inhomogeneous Schrödinger equations (9.40) with the initial conditions $\sigma_n(0) = \sigma'_n(0) = 0$ can be taken in the form (c.f., [108])

$$\sigma_n(x) = f(x) \int_0^x \left(\frac{1}{f^2(s)} \int_0^s f(t) h_n(t) dt \right) ds.$$

Substituting the right-hand side from Eq. (9.36) and performing several integrations by parts to get rid of the derivatives of the function σ_{n-2} under the integral signs, we obtain the following recurrent formulas for the functions σ_n and τ_n:

$$\eta_n(x) = \int_0^x \left(t f'(t) + (n-1) f(t) \right) \sigma_{n-2}(t) \, dt, \quad \theta_n(x) = \int_0^x \frac{1}{f^2(t)} \left(\eta_n(t) - t f(t) \sigma_{n-2}(t) \right) dt,$$

$$\sigma_n(x) = \frac{2n+1}{2n-3} \left[x^2 \sigma_{n-2}(x) + c_n f(x) \theta_n(x) \right], \tag{9.41}$$

$$\tau_n(x) = \frac{2n+1}{2n-3} \left[x^2 \tau_{n-2}(x) + c_n \left(f'(x) \theta_n(x) + \frac{\eta_n(x)}{f(x)} \right) - (c_n - 2n + 1) x \sigma_{n-2}(x) \right], \tag{9.42}$$

for $n = 1, 2, \ldots$, where $c_n = 1$ if $n = 1$ and $c_n = 2(2n - 1)$ otherwise.

Let us explain why the series (9.12) and (9.13) can be differentiated termwise. Suppose that $q \in C^2[0, b]$. First, it follows from the equality

$$\sum_{n=0}^{\infty} (2n + 1) j_n^2(z) = 1$$

([5, 10.1.50]) and the Cauchy–Schwarz inequality that a series

$$\sum_{n=0}^{\infty} a_n(x) j_{2n+\delta}(\rho x)$$

(where δ is zero or one) is uniformly convergent provided that the series

$$\sum_{n=0}^{\infty} \frac{a_n^2(x)}{n}$$

is uniformly convergent. Second, it follows from [81, Corollary I to Theorem XIV] and [169] that if the function $g \in C^{p+1}[-1, 1]$, then its Fourier-Legendre coefficients $a_n(g)$

satisfy the inequality

$$|a_n(g)| \le \frac{c_p V}{n^{p+1/2}},$$

where c_p is a universal constant and $V = \max_{[-1,1]} |g^{(p+1)}(x)|$.

Consider the coefficients β_n. As can be seen from (9.7), $\beta_n \in C^2[0, b]$ (at $x = 0$ we define β_n by continuity). Moreover, $\beta_n(x)$ are the Fourier–Legendre coefficients of the function $x K_h(x, xz) \in C^3[-1, 1]$ with respect to z. Hence, $|\beta_n(x)| \le c_3 n^{-5/2}$. For the derivatives we have

$$\beta_n'(x) = \frac{2n + 1}{2} \int_{-1}^{1} \left(K_h(x, xz) + x K_{h,1}(x, xz) + xz K_{h,2}(x, xz) \right) P_n(z)\, dz,$$

i.e., $\beta_n'(x)$ are the Fourier–Legendre coefficients of the function $K_h(x, xz) + x K_{h,1}(x, xz) + xz K_{h,2}(x, xz) \in C^2[-1, 1]$. Hence $|\beta_n'(x)| \le c_2 n^{-3/2}$. Similarly, $|\beta_n''(x)| \le c_1 n^{-1/2}$, providing the uniform convergence of all series involved in this section.

The validity of the formulas (9.41) and (9.42) in the general case can be verified by taking a sequence $q_n \in C^2[0, b]$ such that $q_n \to q$ uniformly as $n \to \infty$, constructing the corresponding integral kernels and coefficients β_n, γ_n for each q_n, and passing to the limit in the formulas (9.41) and (9.42). Proposition 9.3 now follows by differentiating (9.41) and (9.42).

Let us summarize the results of this section. The coefficients β_k and γ_k, $k = 0, 1, \ldots$ can be efficiently computed in the following way.

1. Compute a nonvanishing solution f of the equation

$$f'' - q(x)f = 0, \quad f(0) = 1$$

   and its derivative f'. In general, f is complex valued, $h := f'(0)$.
2. Define the first coefficients as

$$\beta_{-1} := 1/2, \quad \beta_0 = (f - 1)/2, \quad \gamma_{-1} := \frac{1}{4} \int_0^x q(s)\, ds$$

   and

$$\gamma_0 = \frac{f' - h}{2} - \frac{1}{4} \int_0^x q(s)\, ds.$$

3. The rest of the coefficients are computed recurrently as $\beta_k = \sigma_k / x^k$ and $\gamma_k = \tau_k / x^k$ by the formulas (9.41) and (9.42).

Numerical tests show that this procedure allows one to compute a large number of the coefficients β_k and γ_k with a high accuracy (see [106]). Note that when solving spectral problems one needs to compute these coefficients only once and then use the representations (9.12), (9.13), (9.26) and (9.27) on any domain of the complex ρ-plane, which requires only the computation of the spherical Bessel functions and arithmetic operations. Moreover, the estimates of the type (9.16) and (9.17) guarantee the uniformity of the approximation error with respect to Re ρ.

Remark 9.7 In [118] the NSBF representations for solutions together with the recurrent procedures for computing the corresponding coefficients were obtained also for the general Sturm–Liouville equation (2.13).

In [119] the NSBF representations were obtained for regular solutions of perturbed Bessel equations.

In Part III we apply the obtained NSBF representations (9.12), (9.13), (9.26) and (9.27) to solve direct Sturm–Liouville problems on finite intervals. The Fourier–Legendre series representations for the transmutation kernels from Sect. 9.2 are used in Chaps. 13–15 to solve inverse spectral problems on finite and semi-infinite intervals.

Series Representations for the Kernel $A(x, t)$ and for the Jost Solution

In this chapter we derive, following [59] and [104], a series representation for the Jost solution $e(\rho, x)$ introduced in Chap. 4, explaining all the steps in detail. An analogous representation for the second Jost solution $g(\rho, x)$ (see Chap. 6) is given in Chap. 16, where it is used to solve the inverse scattering problem on the line.

The first step is to obtain a convenient series representation for the kernel $A(x, t)$ from Sect. 8.3. This is done in the next section.

10.1 Fourier–Laguerre Series for $A(x, t)$

Consider the one-dimensional Schrödinger equation

$$- y'' + q(x)y = \lambda y, \quad x > 0, \tag{10.1}$$

where $q(x)$ is a real-valued function that satisfies the condition

$$\int_0^\infty (1 + x)\,|q(x)|\,dx < \infty. \tag{10.2}$$

We are interested in obtaining a series representation for the Jost solution $e(\rho, x)$ starting from its Levin integral representation

$$e(\rho, x) = e^{i\rho x} + \int_x^\infty A(x, t)e^{i\rho t}\,dt. \tag{10.3}$$

V. V. Kravchenko, *Direct and Inverse Sturm-Liouville Problems*, Frontiers in Mathematics, https://doi.org/10.1007/978-3-030-47849-0_10

Denote

$$a(x, t) := e^{\frac{t}{2}} A(x, x + t).$$

Since $A(x, \cdot) \in L_2(x, \infty)$ we have that $a(x, \cdot)$ belongs to $L_2(0, \infty; e^{-t})$. Indeed,

$$\int_0^\infty e^{-t} a^2(x, t) dt = \int_0^\infty A^2(x, x + t) dt = \int_x^\infty A^2(x, t) dt < \infty.$$

Hence, $a(x, t)$ admits the representation (see, e.g., [161])

$$a(x, t) = \sum_{n=0}^\infty a_n(x) L_n(t)$$

where L_n is the *Laguerre polynomial* of order n. For each fixed $x \geq 0$, the series converges in the norm of the space $L_2(0, \infty; e^{-t})$.

Among the notable properties of the Laguerre polynomials we recall that

$$\int_0^\infty e^{-t} L_m(t) L_n(t) dt = \delta_{nm}$$

where δ_{nm} is the Kronecker symbol, and

$$L_n(0) = 1 \quad \text{for all } n = 0, 1, \ldots. \tag{10.4}$$

The system of Laguerre polynomials $\{L_n(t)\}_{n=0}^\infty$ forms an orthonormal basis in the Hilbert space $L_2(0, \infty; e^{-t})$.

Returning to the kernel $A(x, t)$, we obtain (see [104]) the equality

$$A(x, t) = \sum_{n=0}^\infty a_n(x) L_n(t - x) e^{\frac{x-t}{2}}. \tag{10.5}$$

This implies that

$$\sum_{n=0}^\infty a_n(x) = A(x, x) = \frac{1}{2} \int_x^\infty q(t) dt. \tag{10.6}$$

This is due to (8.16) and the property (10.4).

10.2 A Representation for the Jost Solution

Substitution of (10.5) in (10.3) leads to a series representation of the Jost solution,

$$e(\rho, x) = e^{i\rho x}\left(1 + \int_0^\infty A(x, x+t)e^{i\rho t}dt\right)$$

$$= e^{i\rho x}\left(1 + \int_0^\infty a(x, t)e^{-\left(\frac{1}{2}-i\rho\right)t}dt\right)$$

$$= e^{i\rho x}\left(1 + \sum_{n=0}^\infty a_n(x)\int_0^\infty L_n(t)e^{-\left(\frac{1}{2}-i\rho\right)t}dt\right).$$

According to [76, formula 7.414 (2)],

$$\int_0^\infty L_n(t)e^{-\left(\frac{1}{2}-i\rho\right)t}dt = \frac{(-1)^n\left(\frac{1}{2}+i\rho\right)^n}{\left(\frac{1}{2}-i\rho\right)^{n+1}}.$$

Hence

$$e(\rho, x) = e^{i\rho x}\left(1 + \sum_{n=0}^\infty \frac{(-1)^n\left(\frac{1}{2}+i\rho\right)^n}{\left(\frac{1}{2}-i\rho\right)^{n+1}}a_n(x)\right). \tag{10.7}$$

Denote

$$z := \frac{\frac{1}{2}+i\rho}{\frac{1}{2}-i\rho}. \tag{10.8}$$

Notice that this is a Möbius transformation of the upper half-plane of the complex variable ρ onto the unit disc $D = \{z \in \mathbb{C} : |z| \le 1\}$. It maps the point $\rho = 0$ to $z = 1$, the ray $\rho = i\tau, \tau > 0$ (corresponding to $\lambda < 0$) to the interval $(-1, 1)$ in terms of z, and when τ runs from 0 to $+\infty$, z runs from 1 to -1. When ρ (and hence λ) runs from 0 to $+\infty$ along the real line, z runs from 1 to -1 along the upper unit half-circle.

From (10.8) we have

$$i\rho = \frac{z-1}{2(z+1)} \quad \text{and} \quad \frac{1}{2} - i\rho = \frac{1}{z+1}.$$

It is then easy to see that in terms of z the Jost solution (10.7) can be written in the form

$$e(\rho, x) = e^{i\rho x} \left(1 + (z+1) \sum_{n=0}^{\infty} (-1)^n z^n a_n(x) \right). \tag{10.9}$$

Thus, the function $e(\rho, x)e^{-i\rho x}$ is just a power series in the parameter z. The series is convergent in the unit disk of the complex z-plane. For any $x \geq 0$, the series $\sum_{n=0}^{\infty} a_n^2(x)$ converges, because a_n are the Fourier coefficients of a function from $L_2\left(0, \infty; e^{-t}\right)$ with respect to the Laguerre polynomials. Hence, for any $x \geq 0$, the function $e(\rho, x)e^{-i\rho x}$ belongs to the Hardy space of the unit disc $H^2(D)$ as a function of z (see, e.g., [153, Theorem 17.12]).

In the next section we derive a system of equations for the coefficients $a_n(x)$.

10.3   Equations for the Coefficients a_n

Substitution of the expression (10.7) in Eq. (10.1) leads to the equality

$$q(x) \left(1 + \sum_{n=0}^{\infty} \frac{(-1)^n \left(\frac{1}{2} + i\rho \right)^n}{\left(\frac{1}{2} - i\rho \right)^{n+1}} a_n(x) \right)$$

$$= \sum_{n=0}^{\infty} \frac{(-1)^n \left(\frac{1}{2} + i\rho \right)^n}{\left(\frac{1}{2} - i\rho \right)^{n+1}} a_n''(x) + 2i\rho \sum_{n=0}^{\infty} \frac{(-1)^n \left(\frac{1}{2} + i\rho \right)^n}{\left(\frac{1}{2} - i\rho \right)^{n+1}} a_n'(x). \tag{10.10}$$

In terms of the parameter z, (10.10) can be written in the form

$$q(x) + q(x)(z+1) \sum_{n=0}^{\infty} (-1)^n z^n a_n(x)$$

$$= (z+1) \sum_{n=0}^{\infty} (-1)^n z^n a_n''(x) + (z-1) \sum_{n=0}^{\infty} (-1)^n z^n a_n'(x).$$

Equating the terms corresponding to equal powers of z we obtain the equalities

$$a_0'' - a_0' - q a_0 = q, \tag{10.11}$$

$$L a_n - a_n' = L a_{n-1} + a_{n-1}', \quad n = 1, 2, \ldots, \tag{10.12}$$

where $L := \frac{d^2}{dx^2} - q(x)$.

Additionally,

$$a_n(x) \to 0 \text{ when } x \to \infty, \quad n = 0, 1, \dots. \tag{10.13}$$

A non-rigorous explanation of this can be given as follows. From (4.6) it follows that

$$\sum_{n=0}^{\infty} (-1)^n z^n a_n(\infty) = 0$$

for all $z \in D$. Hence, $a_n(\infty) = 0$, $n = 0, 1, \dots$. This argument is applicable whenever the limits of the coefficients a_n at infinity exist. For a_0 this follows directly from (10.7). Indeed, substitution of $\rho = i/2$ in (10.7) leads to the equality

$$a_0(x) = e\left(\frac{i}{2}, x\right) e^{\frac{x}{2}} - 1,$$

which implies that (10.13) holds for a_0 due to the asymptotics of the Jost solution at infinity (4.6). Below we prove (10.13) for the rest of the coefficients and show how they can be constructed. Moreover, we prove the uniqueness of the solutions a_n, $n = 0, 1, \dots$ of the system (10.11), (10.12) satisfying (10.13).

Proposition 10.1 *In terms of the functions* $\beta_n := 1 + a_n$, *the system (10.11) and (10.12) can be written as*

$$L\beta_0 - \beta_0' = 0, \tag{10.14}$$

$$L\beta_n - \beta_n' = 2 \sum_{k=0}^{n-1} \beta_k', \quad n = 1, 2, \dots. \tag{10.15}$$

Proof Equation (10.14) is obvious. We prove (10.15) by induction. For a_1 relation (10.12) gives $La_1 - a_1' = La_0 + a_0'$, where due to (10.11), $La_0 = q + a_0'$. Hence $La_1 - a_1' = q + 2a_0'$ and in terms of β_1: $L\beta_1 - \beta_1' = 2a_0'$, and thus for $n = 1$ Eq. (10.15) is proved. Now assuming that (10.15) is valid for $n - 1$, relation (10.12) gives that

$$La_n - a_n' = La_{n-1} + a_{n-1}' = q + 2 \sum_{k=0}^{n-1} a_k'$$

which in terms of β_n gives (10.15). ∎

Theorem 10.1 *Under the condition (10.2), the transmutation kernel $A(x,t)$ admits the representation (10.5), where for each fixed x the series converges in the norm of $L_2(x,\infty)$ and the coefficients a_n satisfy the Eqs. (10.11), (10.12) and (10.13).*

The Jost solution admits the representation (10.7), or (10.9) if written in terms of z.

Proof Let us show that the system of equations (10.11) and (10.12) admits solutions satisfying (10.13) and such solutions are unique.

Notice that the differential expressions in the left-hand side of (10.14) and (10.15) can be factorized as

$$\left(L - \frac{d}{dx}\right)\beta_n = e^{\frac{x}{2}}\left(\frac{d^2}{dx^2} - \left(\frac{1}{4} + q(x)\right)\right)\left(e^{-\frac{x}{2}}\beta_n\right).$$

Thus, for the functions $\zeta_n(x) := e^{-\frac{x}{2}}\beta_n(x)$ we obtain the equations

$$\left(\frac{d^2}{dx^2} - \left(\frac{1}{4} + q(x)\right)\right)\zeta_0(x) = 0, \tag{10.16}$$

$$\left(\frac{d^2}{dx^2} - \left(\frac{1}{4} + q(x)\right)\right)\zeta_n(x) = 2e^{-\frac{x}{2}}\sum_{k=0}^{n-1}\beta'_k \quad n = 1, 2, \dots. \tag{10.17}$$

Since $\zeta_n(x) = e^{-\frac{x}{2}}(1 + a_n(x))$, to prove the existence and uniqueness of the solutions a_n of the system (10.11), (10.12) satisfying (10.13) is equivalent to proving that there exist the unique solutions ζ_n of the system (10.16) and (10.17) with the asymptotics

$$\zeta_n(x) = e^{-\frac{x}{2}}(1 + o(1)), \quad x \to \infty, \quad n = 0, 1, \dots. \tag{10.18}$$

Indeed, if (10.13) hold, then (10.18) follow immediately from the definition of ζ_n. On the other hand, if (10.18) hold, then

$$a_n(x) = \zeta_n(x)e^{\frac{x}{2}} - 1 = (1 + o(1)) - 1 = o(1), \quad x \to \infty.$$

We prove the existence and the uniqueness of the solutions ζ_n satisfying (10.18), and thus of a_n satisfying (10.13), by induction. First of all, notice that $\zeta_0(x) := e\left(\frac{i}{2}, x\right)$ is the unique solution of (10.16) satisfying (10.18). Hence, $a_0(x) = e\left(\frac{i}{2}, x\right)e^{\frac{x}{2}} - 1$.

Next, assume that functions $a_1, \dots, a_{n-1}$ satisfying (10.13) exist and consider Eqs. (10.17).

A particular solution of (10.17) can be constructed in the form

$$\zeta_{n,0}(x) = -\int_x^\infty \begin{vmatrix} y_1(t) & y_2(t) \\ y_1(x) & y_2(x) \end{vmatrix} \frac{f(t)}{W(t)} dt,$$

where f is the right-hand side of the equation, the functions y_1 and y_2 represent a fundamental system of solutions of the corresponding homogeneous equation and $W(t)$ is their Wronskian.

Denote by $\eta(x)$ a solution of (10.16) constructed with the aid of Abel's formula,

$$\eta(x) := e\left(\frac{i}{2}, x\right) \int_0^x \frac{dt}{e^2\left(\frac{i}{2}, t\right)}. \tag{10.19}$$

Then it is easy to verify (see, e.g., [66, p. 397]) that its asymptotics is $\eta(x) = e^{\frac{x}{2}}(1 + o(1))$, $x \to \infty$. Choosing $y_1(x) = e\left(\frac{i}{2}, x\right)$ and $y_2(x) = \eta(x)$, we have that $W(t) \equiv 1$, and hence

$$\zeta_{n,0}(x) = -2\int_x^\infty \begin{vmatrix} e\left(\frac{i}{2}, t\right) & \eta(t) \\ e\left(\frac{i}{2}, x\right) & \eta(x) \end{vmatrix} e^{-\frac{t}{2}} \sum_{k=0}^{n-1} \beta_k'(t)dt$$

$$= -2\left(\eta(x)J_{1,n}(x) - e\left(\frac{i}{2}, x\right)J_{2,n}(x)\right), \tag{10.20}$$

where

$$J_{1,n}(x) := \int_x^\infty e\left(\frac{i}{2}, t\right) e^{-\frac{t}{2}} \sum_{k=0}^{n-1} \beta_k'(t)dt$$

and

$$J_{2,n}(x) := \int_x^\infty \eta(t)e^{-\frac{t}{2}} \sum_{k=0}^{n-1} \beta_k'(t)dt.$$

Taking into account the asymptotics of $\eta(t)$ at infinity, we have that

$$J_{2,n}(x) = \int_x^\infty (1 + o(1)) \sum_{k=0}^{n-1} \beta_k'(t)dt.$$

Since, due to our assumption,

$$\int_x^\infty \beta_k'(t)dt = \int_x^\infty a_k'(t)dt = a_k(\infty) - a_k(x) = -a_k(x), \quad k = 0, \ldots, n-1,$$

we conclude that $J_{2,n}(x) = o(1)$, $x \to \infty$.

Integration by parts gives us the asymptotics of $J_{1,n}(x)$:

$$J_{1,n}(x) = -e^{-\frac{x}{2}} e\left(\frac{i}{2}, x\right) \sum_{k=0}^{n-1} \beta_k(x) + \frac{1}{2} \int_x^\infty e\left(\frac{i}{2}, t\right) e^{-\frac{t}{2}} \sum_{k=0}^{n-1} \beta_k(t) dt$$

$$- \int_x^\infty e'\left(\frac{i}{2}, t\right) e^{-\frac{t}{2}} \sum_{k=0}^{n-1} \beta_k(t) dt$$

$$= -e^{-x}(1 + o(1)) + \frac{1}{2} e^{-x}(1 + o(1)) + \frac{1}{2} e^{-x}(1 + o(1))$$

$$= e^{-x} o(1), \quad x \to \infty.$$

Substituting the asymptotics of $J_{1,n}(x)$ and $J_{2,n}(x)$ in (10.20) we get $\zeta_{n,0}(x) = e^{-\frac{x}{2}} o(1)$. Hence, the unique solution ζ_n of (10.17), such that $\zeta_n(x) = e^{-\frac{x}{2}}(1 + o(1))$, $x \to \infty$, has the form $\zeta_n(x) = e\left(\frac{i}{2}, x\right) + \zeta_{n,0}(x)$, where $\zeta_{n,0}(x)$ has the form (10.20). ■

10.4   Estimates for the Representation of the Jost Solution

Together with the exact series representation (10.7) for the Jost solution, consider its approximation

$$e_N(\rho, x) := e^{i\rho x}\left(1 + \sum_{n=0}^N \frac{(-1)^n \left(\frac{1}{2} + i\rho\right)^n}{\left(\frac{1}{2} - i\rho\right)^{n+1}} a_n(x)\right).$$

The following estimates of its accuracy are valid.

Theorem 10.2

(1) *Let* $\operatorname{Im} \rho > 0$. *Then*

$$|e(\rho, x) - e_N(\rho, x)| \leq \varepsilon_N(x) \frac{e^{-\operatorname{Im}\rho x}}{\sqrt{2\operatorname{Im}\rho}}, \tag{10.21}$$

where

$$\varepsilon_N(x) := \left(\sum_{N+1}^\infty |a_n(x)|^2\right)^{1/2} = \left(\int_0^\infty e^{-t} |a(x, t) - a_N(x, t)|^2 dt\right)^{1/2}, \tag{10.22}$$

with $a_N(x, t) := \sum_{n=0}^N a_n(x) L_n(t)$ *(equality (10.22) follows from Parseval's identity).*

(2) *For $\rho \in \mathbb{R}$,*

$$\|e(\cdot, x) - e_N(\cdot, x)\|_{L_2(-\infty,\infty)} = \sqrt{2\pi}\,\varepsilon_N(x). \qquad (10.23)$$

Proof

(1) Assume $\mathrm{Im}\,\rho > 0$ and consider

$$|e(\rho, x) - e_N(\rho, x)| = \left| \int_0^\infty e^{-t}(a(x,t) - a_N(x,t))e^{\frac{t}{2}+i\rho(x+t)}dt \right|$$

$$= \left| \left\langle \bar{a}(x,t) - \bar{a}_N(x,t), e^{\frac{t}{2}+i\rho(x+t)} \right\rangle \right|_{L_2(0,\infty;e^{-t})}.$$

The Cauchy–Bunyakovsky–Schwarz inequality implies that

$$|e(\rho, x) - e_N(\rho, x)| \le \varepsilon_N(x) \left\| e^{\frac{t}{2}+i\rho(x+t)} \right\|_{L_2(0,\infty;e^{-t})}.$$

We have

$$\left\| e^{\frac{t}{2}+i\rho(x+t)} \right\|_{L_2(0,\infty;e^{-t})} = \left(\int_0^\infty e^{-2\,\mathrm{Im}\,\rho(x+t)}dt \right)^{1/2} = \frac{e^{-\mathrm{Im}\,\rho\,x}}{\sqrt{2\,\mathrm{Im}\,\rho}},$$

and so (10.21) is proved.

(2) Consider the difference

$$e(\rho, x) - e_N(\rho, x) = \int_x^\infty (A(x,t) - A_N(x,t))\,e^{i\rho t}dt.$$

Extending the function $A(x,t) - A_N(x,t)$ by zero for $t \in (-\infty, x)$, we obtain that the function $e(\rho, x) - e_N(\rho, x)$ is a Fourier transform of an L_2-function, and hence

$$\|e(\cdot, x) - e_N(\cdot, x)\|_{L_2(-\infty,\infty)} = \sqrt{2\pi}\,\|A(x,\cdot) - A_N(x,\cdot)\|_{L_2(-\infty,\infty)}.$$

Note that

$$A(x,t) - A_N(x,t) = \sum_{N+1}^\infty a_n(x)L_n(t-x)e^{\frac{x-t}{2}}$$

and hence

$$\|A(x,\cdot) - A_N(x,\cdot)\|^2_{L_2(-\infty,\infty)} = \int_0^\infty \left|\sum_{N+1}^\infty a_n(x)L_n(t)\right|^2 e^{-t}dt = \varepsilon_N^2(x)$$

which gives (10.23).

$\blacksquare$

10.5   A Recurrent Integration Procedure for Computing the Coefficients a_n

The proof of Theorem 10.1 gives us a method for calculating the coefficients a_n.

1. Compute the Jost solution $e\left(\frac{i}{2},x\right)$ together with the second solution $\eta(x)$ of (10.16) and their derivatives. To find an approximation $\tilde{e}\left(\frac{i}{2},x\right)$ of $e\left(\frac{i}{2},x\right)$ one can compute a solution of the Cauchy problem for (10.16) on a sufficiently large interval $(0,b)$, with the initial conditions

$$\tilde{e}\left(\frac{i}{2},b\right) = e^{-\frac{b}{2}} \quad \text{and} \quad \tilde{e}'\left(\frac{i}{2},b\right) = -\frac{e^{-\frac{b}{2}}}{2}$$

which follow from the asymptotics of the Jost solution.
    Next, $\eta(x)$ is computed by (10.19).
2. Compute $a_0(x) = e^{\frac{x}{2}}e\left(\frac{i}{2},x\right) - 1$.
3. Compute $a_n(x) = a_0(x) - 2e^{\frac{x}{2}}\left(\eta(x)J_{1,n}(x) - e\left(\frac{i}{2},x\right)J_{2,n}(x)\right)$.

To compute $J_{1,n}$ and $J_{2,n}$ we can adopt the following strategy. Integration by parts leads to the formulas

$$J_{1,n}(x) = -e^{-\frac{x}{2}}e\left(\frac{i}{2},x\right)\sum_{k=0}^{n-1}a_k(x) - \int_x^\infty \left(e\left(\frac{i}{2},t\right)e^{-\frac{t}{2}}\right)'\sum_{k=0}^{n-1}a_k(t)dt$$

and

$$J_{2,n}(x) = -e^{-\frac{x}{2}}\eta(x)\sum_{k=0}^{n-1}a_k(x) - \int_x^\infty \left(\eta(t)e^{-\frac{t}{2}}\right)'\sum_{k=0}^{n-1}a_k(t)dt.$$

It follows that

$$J_{1,n}(x) = J_{1,n-1}(x) - e^{-\frac{x}{2}} e\left(\frac{i}{2}, x\right) a_{n-1}(x) - \int_x^\infty \left(e\left(\frac{i}{2}, t\right) e^{-\frac{t}{2}}\right)' a_{n-1}(t)dt$$

$$(10.24)$$

and

$$J_{2,n}(x) = J_{2,n-1}(x) - e^{-\frac{x}{2}} \eta(x) a_{n-1}(x) - \int_x^\infty \left(\eta(t) e^{-\frac{t}{2}}\right)' a_{n-1}(t)dt. \qquad (10.25)$$

10.6   A Representation for the Derivative of the Jost Solution

In this section we additionally assume that q is an absolutely continuous function. In order to obtain a series representation for the derivative of the Jost solution for which an analogue of Theorem 10.2 would be valid, we begin again with (10.3). Its differentiation gives

$$e'(\rho, x) = i\rho e^{i\rho x} - A(x, x)e^{i\rho x} + \int_x^\infty A_x(x, t)e^{i\rho t}dt$$

$$= i\rho e^{i\rho x}\left(1 + \frac{\omega(x)}{i\rho}\right) + \int_x^\infty A_x(x, t)e^{i\rho t}dt, \qquad (10.26)$$

where

$$\omega(x) := -\frac{1}{2}\int_x^\infty q(t)dt.$$

The derivative $A_x(x, t)$ exists and $A_x(x, \cdot) \in L_1(x, \infty)$ [127, Chapter 1]. Moreover [172, p. 136], $A_x(x, \cdot)$ is absolutely continuous and hence belongs to $L_2(x, \infty)$. Thus, we can argue for the kernel $A_x(x, t)$ the same way we did for $A(x, t)$. Denote

$$a_1(x, t) := e^{\frac{t}{2}} A_x(x, x + t).$$

This function admits a Fourier–Laguerre series representation

$$a_1(x, t) = \sum_{n=0}^\infty d_n(x)L_n(t).$$

Then

$$A_x(x, t) = \sum_{n=0}^{\infty} d_n(x) L_n(t - x) e^{\frac{x-t}{2}}.$$

Substitution of this expression in (10.26) leads, similarly to Sect. 10.2, to the following series representation for the derivative of the Jost solution:

$$e'(\rho, x) = e^{i\rho x} \left(i\rho + \omega(x) + \sum_{n=0}^{\infty} \frac{(-1)^n \left(\frac{1}{2} + i\rho \right)^n}{\left(\frac{1}{2} - i\rho \right)^{n+1}} d_n(x) \right).$$

For this representation we can derive estimates similar to those from Theorem 10.2. We omit this step.

In terms of z we have

$$e'(\rho, x) = e^{i\rho x} \left(\frac{z - 1}{2(z + 1)} + \omega(x) + (z + 1) \sum_{n=0}^{\infty} (-1)^n z^n d_n(x) \right). \tag{10.27}$$

On the other hand, (10.7) directly yields

$$e'(\rho, x) = i\rho e^{i\rho x} \left(1 + \sum_{n=0}^{\infty} \frac{(-1)^n \left(\frac{1}{2} + i\rho \right)^n}{\left(\frac{1}{2} - i\rho \right)^{n+1}} a_n(x) \right)$$

$$+ e^{i\rho x} \sum_{n=0}^{\infty} \frac{(-1)^n \left(\frac{1}{2} + i\rho \right)^n}{\left(\frac{1}{2} - i\rho \right)^{n+1}} a'_n(x)$$

or, in terms of z,

$$e'(\rho, x) = e^{i\rho x} \frac{z - 1}{2(z + 1)} \left(1 + (z + 1) \sum_{n=0}^{\infty} (-1)^n z^n a_n(x) \right)$$

$$+ e^{i\rho x} (z + 1) \sum_{n=0}^{\infty} (-1)^n z^n a'_n(x). \tag{10.28}$$

Equating (10.27) with (10.28) we obtain the following equality of power series for all $z \in D$:

$$\omega(x) + (z+1) \sum_{n=0}^{\infty} (-1)^n z^n d_n(x) = \frac{(z-1)}{2} \sum_{n-0}^{\infty} (-1)^n z^n a_n(x)$$

$$+ (z+1) \sum_{n=0}^{\infty} (-1)^n z^n a_n'(x).$$

Thus,

$$d_0(x) = a_0'(x) - \frac{a_0(x)}{2} - \omega(x) \tag{10.29}$$

and

$$d_{n+1}(x) = d_n(x) + a_{n+1}'(x) - a_n'(x) - \frac{1}{2} \left(a_{n+1}(x) + a_n(x) \right), \quad n = 0, 1, \dots. \tag{10.30}$$

Hence to compute the coefficients d_n it will be convenient to have recurrent formulas for $a_n'(x)$. These can be obtained with the aid of formulas from Sect. 10.5. Thus,

$$a_0'(x) = \left(e^{\frac{x}{2}} e\left(\frac{i}{2}, x\right) \right)', \tag{10.31}$$

$$a_n'(x) = a_0'(x) - 2 \left(e^{\frac{x}{2}} \eta(x) \right)' J_{1,n}(x) - 2 e^{\frac{x}{2}} \eta(x) J_{1,n}'(x)$$

$$+ 2 \left(e^{\frac{x}{2}} e\left(\frac{i}{2}, x\right) \right)' J_{2,n}(x) + 2 e^{\frac{x}{2}} e\left(\frac{i}{2}, x\right) J_{2,n}'(x), \tag{10.32}$$

where by (10.24) and (10.25) the following relations hold for $J_{1,n}'(x)$ and $J_{2,n}'(x)$:

$$J_{1,n}'(x) = J_{1,n-1}'(x) - e^{-\frac{x}{2}} e\left(\frac{i}{2}, x\right) a_{n-1}'(x) \tag{10.33}$$

and

$$J_{2,n}'(x) = J_{2,n-1}'(x) - e^{-\frac{x}{2}} \eta(x) a_{n-1}'(x). \tag{10.34}$$

Remark 10.1 Notice that

$$A_x(x, x) = \sum_{n=0}^{\infty} d_n(x)$$

while from (8.16) we have that

$$\frac{d}{dx} A(x, x) = -\frac{q(x)}{2}.$$

Moreover, differentiating (10.5) with respect to t and substituting $t = x$ yields

$$A_t(x, x) = -\sum_{n=0}^{\infty} \left(n + \frac{1}{2}\right) a_n(x).$$

Since $A_x(x, x) = \frac{d}{dx} A(x, x) - A_t(x, x)$, we obtain the equality

$$\sum_{n=0}^{\infty} d_n(x) = -\frac{q(x)}{2} + \sum_{n=0}^{\infty} \left(n + \frac{1}{2}\right) a_n(x). \tag{10.35}$$

The formulas derived in the last two sections for computing the coefficients a_n and d_n, $n = 0, 1, \ldots$ are quite simple and convenient for numerical implementation. Again, similarly to the NSBF representations for the solutions $\varphi(\rho, x)$ and $s(\rho, x)$, an attractive feature of the representation (10.9) for the Jost solution and (10.27) for its derivative is that the coefficients a_n and d_n, $n = 0, 1, \ldots$ need to be computed only once for a chosen interval in x, and then the representations can be used for solving spectral problems by calculating partial sums of (10.9) and (10.27) on any subset of the unit disk D in terms of the parameter z. As we explain in Chap. 12, this leads to a simple, practical and powerful method for solving Sturm–Liouville problems with short-range potentials on infinite intervals.

Part III

Solution of Direct Sturm-Liouville Problems

In this part the functional series representations for solutions of Sturm-Liouville equations obtained in the preceding chapters are used for solving direct Sturm-Liouville problems on finite and infinite intervals. We begin with the SPPS method for solving problems on finite intervals. Its main advantages are the simplicity for numerical implementation and adaptability to different boundary conditions and models. Next we consider the NSBF method and show that it is a powerful tool for computing very large sets of eigendata quickly and accurately. Finally, taking as an example the Sturm-Liouville problem on a half-line, we present an approach based on the functional series representation for the Jost solution, derived in Chap. 10. We show that with the aid of the results of Chap. 10 this computationally difficult problem turns into an easy exercise. Other spectral problems on infinite intervals can be solved by implementing the described technique.

Sturm–Liouville Problems on Finite Intervals   **11**

In this chapter, following [108] and [106], we briefly explain how the SPPS and NSBF representations from Sect. 2.2 and Chap. 9, respectively, can be used for solving the Sturm–Liouville problem on a finite interval.

11.1   The SPPS Method

The fact that Sturm Liouville problems are related to the problem of finding zeros of complex analytic functions of the variable λ is quite well known (see, e.g., [128]). For a regular Sturm–Liouville problem the corresponding analytic function is even entire. The SPPS representation (Theorem 2.1) allows us to obtain the Taylor series of that analytic function explicitly. As an example, consider first a spectral problem for the equation

$$-y'' + q(x)y = \lambda y, \quad 0 < x < 1 \tag{11.1}$$

with the boundary conditions

$$y(0) = 0 \quad \text{and} \quad y(1) = 0. \tag{11.2}$$

Assume that the coefficient q satisfies the conditions from Remark 2.6 and that f is constructed as described there, taking $x_0 = 0$. The first boundary condition and Remark 2.1 imply that the constant c_1 in (2.9) must be zero. Then the spectral problem reduces to

V. V. Kravchenko, *Direct and Inverse Sturm-Liouville Problems*,
Frontiers in Mathematics, https://doi.org/10.1007/978-3-030-47849-0_11

finding values of λ for which

$$y_2(1) = \sum_{k=0}^{\infty} \frac{(-\lambda)^k}{(2k+1)!} \varphi_{2k+1}(1)$$

(see (2.10)) vanishes. In other words, this spectral problem reduces to the calculation of the zeros of the complex analytic function $\kappa(\lambda) = \sum_{m=0}^{\infty} a_m \lambda^m$, where

$$a_m = (-1)^m \frac{\varphi_{2m+1}(1)}{(2m+1)!}.$$

Now let α and β be arbitrary real numbers and consider the more general boundary conditions

$$y(a)\cos\alpha + y'(a)\sin\alpha = 0, \tag{11.3}$$

$$y(b)\cos\beta + y'(b)\sin\beta = 0 \tag{11.4}$$

together with the equation

$$-y'' + q(x)y = \lambda y, \quad a < x < b. \tag{11.5}$$

Taking the solutions y_1 and y_2 defined by (2.10) and using Remark 2.1 with $x_0 = a$, we obtain the following equation,

$$c_1(f(a)\cos\alpha + f'(a)\sin\alpha) + c_2 \frac{\sin\alpha}{f(a)} = 0,$$

which gives $c_2 = \gamma c_1$ when $\alpha \neq \pi n$, with $\gamma = -f(a)(f(a)\cot\alpha + f'(a))$, and $c_1 = 0$ when $\alpha = \pi n$. In the latter case the result is similar to the example considered above, so let us take $\alpha \neq \pi n$.

The boundary condition (11.4) implies that

$$\left(f(b)\cos\beta + f'(b)\sin\beta\right) \left(\sum_{k=0}^{\infty} (-\lambda)^k \frac{\tilde{X}^{(2k)}(b)}{(2k)!} + \gamma \sum_{k=0}^{\infty} (-\lambda)^k \frac{X^{(2k+1)}(b)}{(2k+1)!}\right)$$

$$+ \frac{\sin\beta}{f(b)} \left(\sum_{k=1}^{\infty} (-\lambda)^k \frac{\tilde{X}^{(2k-1)}(b)}{(2k-1)!} + \gamma \sum_{k=0}^{\infty} (-\lambda)^k \frac{X^{(2k)}(b)}{(2k)!}\right) = 0.$$

Thus the spectral problem (11.5), (11.3), and (11.4) reduces to the problem of calculating the zeros of the analytic function $\kappa(\lambda) = \sum_{m=0}^{\infty} a_m (-\lambda)^m$, where

$$a_0 = \left(f(b)\cos\beta + f'(b)\sin\beta\right)(1 + \gamma X^{(1)}(b)) + \frac{\gamma \sin\beta}{f(b)}$$

and

$$a_m = \left(f(b)\cos\beta + f'(b)\sin\beta\right)\left(\frac{\widetilde{X}^{(2m)}(b)}{(2m)!} + \gamma \frac{X^{(2m+1)}(b)}{(2m+1)!}\right)$$

$$+ \frac{\sin\beta}{f(b)}\left(\frac{\widetilde{X}^{(2m-1)}(b)}{(2m-1)!} + \gamma \frac{X^{(2m)}(b)}{(2m)!}\right), \quad m = 1, 2, \ldots.$$

This reduction of a Sturm–Liouville spectral problem lends itself to a simple numerical implementation. To calculate the first n eigenvalues, we consider the Taylor polynomial $\kappa_N(\lambda) = \sum_{m=0}^{N} a_m (-\lambda)^m$ with $N \geq n$. Thus the numerical approximation of eigenvalues of the Sturm–Liouville problem reduces to the calculation of zeros of the polynomial $\kappa_N(\lambda)$.

There is no need to work with zeros of only one polynomial. It is well known that in general the higher roots of a polynomial become less stable with respect to small inaccuracies in coefficients. Our spectral parameter power series method is well suited to overcome this problem and thus to calculate higher eigenvalues with a good accuracy. This is done using Remark 2.5. Suppose we have already calculated the eigenvalue λ_0 using the procedure described above as a first root of the obtained polynomial. Then for the next step we set $Y_0 = y_1 + iy_2$, where y_1 and y_2 are defined by (2.10) with $\lambda = \lambda_0$. The function Y_0 is then a solution of (2.12). We use it to obtain the eigenvalue λ_1 of the original problem observing that $\lambda_1 = \Lambda_1 + \lambda_0$, where Λ_1 is the first eigenvalue of the equation $(L - \lambda_0)y = \Lambda y$ with the same boundary conditions as in the original problem. This procedure can be continued for calculating higher eigenvalues. Note that if $\lambda_0 = 0$ we should begin this shifting procedure starting with λ_1.

Here we discuss some numerical examples from [108].

Paine Problem A number of spectral problems which have become standard test cases appear in [138] and [142]. As a first example, consider

$$q(x) = \frac{1}{(x + 0.1)^2},$$

$$y(0) = 0, \quad y(\pi) = 0.$$

The eigenvalues displayed in the following table were calculated via SPPS using integration on 10,000 subintervals to calculate $N = 100$ powers of λ. These eigenvalues were found as roots of a single polynomial (i.e., the shifting of λ as described in Remark 2.5 was not applied). Due to the sensitivity of the larger roots of the polynomial to errors in the coefficients, 100-digit arithmetic was used.

n	λ_n [142]	λ_n SPPS
0	1.5198658211	1.519865821099
1	4.9433098221	4.943309822144
2	10.284662645	10.28466264509
3	17.559957746	17.55995774633
4	26.782863158	26.78286315899
5	37.964425862	37.96442587941
6	51.113357757	51.11335707578
7	66.236447704	66.23646092491
8	83.338962374	83.33879073183
9	102.42498840	102.4259718823
10	123.49770680	123.512483827

On the basis of the above values, a new calculation was made by shifting with $\lambda^* = 66$, resulting in the following improved approximations for the last few eigenvalues.

n	λ_n [142]	λ_n SPPS
7	66.236447704	66.23644770359
8	83.338962374	83.33896237419
9	102.42498840	102.42498839828
10	123.49770680	123.49770680101
11	146.55960608	146.55960605783
12	171.61264485	171.61265439928

With $\lambda^* = 146$ and increasing the number of powers to $N = 150$, the following further values were obtained.

n	λ_n [142]	λ_n SPPS
11	146.55960608	146.55586199495330
12	171.61264485	171.60875781110985
13	198.65837500	198.65416389844202

When the number of digits for internal calculations was increased to 150, SPPS produced the same results.

Coffey–Evans Equation   This test case, defined by

$$q(x) = -2\beta \cos 2x + \beta^2 \sin^2 2x.$$

$$u(-\pi/2) = 0, \quad u(\pi/2) = 0,$$

presents the challenge of distinguishing eigenvalues within the triple clusters which form as the parameter β increases. We present results for $\beta = 20, 30, 50$. In all cases given here the eigenvalues were obtained without shifting λ.

$\beta = 20$	
$M = 10{,}000$ subintervals, $N = 180$ powers, 100 digits of precision	
n \ λ_n [54, 124]	λ_n SPPS
0 \ −0.00000000000000	0.0000000000000003
1 \ 77.91619567714397	77.9161956771439703
2 \ 151.46277834645663	151.4627783464566396
3 \ 151.46322365765863	151.4632236576586490
4 \ 151.46366898835165	151.4636689883516575
5 \ 220.15422983525995	220.1542298352599497
6 \ 283.0948	283.0948146954014377
7 \ 283.2507	283.2507437431126800
8 \ 283.4087	283.4087354034293064

$\beta = 30$	
$M = 10{,}000$ subintervals, $N = 150$ powers, 100 digits of precision	
n \ λ_n [124, 142]	λ_n SPPS
0 \ 0.0000000000000	0.000000000000000002
1 \ 117.946307662070	117.94630766206876
2 \	231.664928928423790
3 \ 231.66492931296	231.664928928423791
4 \	231.664930082035462
5 \	340.888299091685489
6 \	403.219684016171863
7 \	403.219684016171917

$\beta = 50$		
$M = 10,000$ subintervals, $N = 150$ powers, 100 digits of precision		
n	λ_n [142]	λ_n SPPS
0	0.00000000000000	0.000000000000000000003
1	197.968726516507	197.96872651650729
2		391.807
3	391.80819148905	391.810
4		547.1397060

11.1.1 Sturm–Liouville Problems with Spectral Parameter Dependent Boundary Conditions

In this subsection we consider Sturm–Liouville problems of the form

$$-y'' + qy = \lambda y, \quad x \in (a, b), \tag{11.6}$$

$$y(a)\cos\alpha + y'(a)\sin\alpha = 0, \quad \alpha \in [0, \pi), \tag{11.7}$$

$$\beta_1 y(b) - \beta_2 y'(b) = \phi(\lambda)\left(\beta_1' y(b) - \beta_2' y'(b)\right), \tag{11.8}$$

where ϕ is a complex-valued function of the variable λ and β_1, β_2, β_1', β_2' are complex numbers. This kind of problem arises in many physical applications (we refer to [30] and references therein) and has been studied in a considerable number of publications [30, 52, 55, 57, 70, 168]. For some special forms of the function ϕ, such as $\phi(\lambda) = \lambda$ or $\phi(\lambda) = \lambda^2 + c_1\lambda + c_2$, results were obtained [55, 168] concerning the regularity of the problem (11.6)–(11.8); we will not dwell upon the details. Our purpose is to show the applicability of the method of spectral parameter power series (SPPS) to this type of Sturm–Liouville problems. For simplicity, let us suppose that $\alpha = 0$ and hence the condition (11.7) becomes $y(a) = 0$. Then as was shown in the preceding section, if an eigenfunction exists, it necessarily coincides with y_2 up to a multiplicative constant.

In this case condition (11.8) becomes equivalent to the equality

$$\left(y_0(b)\phi_1(\lambda) - y_0'(b)\phi_2(\lambda)\right)\sum_{k=0}^{\infty}(-\lambda)^k\frac{X^{(2k+1)}(b)}{(2k+1)!} + \frac{\phi_2(\lambda)}{y_0(b)}\sum_{k=0}^{\infty}(-\lambda)^k\frac{X^{(2k)}(b)}{(2k)!} = 0 \tag{11.9}$$

where $\phi_{1,2}(\lambda) = \beta_{1,2} - \beta_{1,2}'\phi(\lambda)$. Calculation of eigenvalues given by (11.9) is especially simple when ϕ is a polynomial of λ. Precisely this particular situation was considered in all of the above mentioned references concerning Sturm–Liouville problems with spectral-parameter dependent boundary conditions. For these problems the calculation of eigenvalues using the SPPS method presents no additional difficulties compared to the parameter-independent situation discussed in the preceding section.

11.2 The NSBF Method

The representations for solutions and their derivatives (9.12), (9.13) and (9.26), (9.27) can be used to solve Eq. (9.11) numerically, and, in particular, to solve numerically related spectral problems. As an example, consider the Sturm–Liouville problem for (11.6) with $a = 0$,

$$\alpha_0 y(0) + \mu_0 y'(0) = 0, \tag{11.10}$$

$$\alpha_b y(b) + \mu_b y'(b) = 0, \tag{11.11}$$

where we allow the coefficients α_0, μ_0, α_b and μ_b to be not only constants, but also entire functions of the square root ρ of the spectral parameter λ satisfying $|\alpha_0| + |\mu_0| \neq 0$ and $|\alpha_b| + |\mu_b| \neq 0$ (for every λ).

Based on the results of Chap. 9 and taking into account that the solutions $\varphi(\rho, x)$ and $s(\rho, x)$ satisfy the initial conditions

$$\varphi(\rho, 0) = 1, \qquad s(\rho, 0) = 0,$$

$$\varphi'(\rho, 0) = h, \qquad s'(\rho, 0) = \rho,$$

we can formulate the following algorithm for solving spectral problems (11.10) and (11.11) for the equation

$$-y'' + q(x)y = \lambda y, \quad 0 < x < b. \tag{11.12}$$

1. Find a solution f of Eq. (2.8) that does not vanish on $[0, b]$. Let f be normalized as $f(0) = 1$ and define $h := f'(0)$. The solution f can be constructed using the SPPS representation (see, e.g., [108] for details) or using any other numerical method. The choice of a particular f does not affect a lot the final accuracy as long as the two functions f and $1/f$ do not take too large values.
2. Compute the functions β_k and γ_k, $k = 0, \ldots, N$ using (9.41) and (9.42).
3. Calculate the approximations $\varphi_N(\rho, x)$ and $s_N(\rho, x)$ of the solutions $\varphi(\rho, x)$ and $s(\rho, x)$ by (9.14) and (9.15). If necessary, calculate the approximations of the derivatives of the solutions using (9.28) and (9.29).

4. The eigenvalues of the problem (11.12), (11.10), and (11.11) coincide with the squares of the zeros of the entire function

$$\Phi(\rho) := \alpha_b \left(\mu_0 \varphi(\rho, b) - (\alpha_0 + \mu_0 h) \frac{s(\rho, b)}{\rho} \right)$$
$$+ \mu_b \left(\mu_0 \varphi'(\rho, b) - (\alpha_0 + \mu_0 h) \frac{s'(\rho, b)}{\rho} \right) \tag{11.13}$$

and are approximated by squares of zeros of the function

$$\Phi_N(\rho) := \alpha_b \left(\mu_0 \varphi_N(\rho, b) - (\alpha_0 + \mu_0 h) \frac{s_N(\rho, b)}{\rho} \right)$$
$$+ \mu_b \left(\mu_0 \overset{\circ}{\varphi}_N(\rho, b) - (\alpha_0 + \mu_0 h) \frac{\overset{\circ}{s}_N(\rho, b)}{\rho} \right). \tag{11.14}$$

5. The eigenfunction y_λ corresponding to the eigenvalue $\lambda = \rho^2$ can be taken in the form

$$y_\lambda = \mu_0 \varphi(\rho, x) - (\alpha_0 + \mu_0 h) \frac{s(\rho, x)}{\rho}. \tag{11.15}$$

Hence once the eigenvalues are calculated, the computation of the corresponding eigenfunctions can be performed using formulas (9.14) and (9.15).

We do not provide the details of the numerical realization of the algorithm, referring the interested reader to [106]. All calculations were performed in Matlab, and for the integrations involved the Newton–Cottes 6-point integration rule was used.

The following observation can be used to estimate an optimal number N to choose.

Remark 11.1 The boundary conditions (8.6) offer a simple and efficient way for controlling the accuracy of the numerical method. Indeed, substitution of (9.6) in (8.6) leads to the equalities

$$\sum_{j=0}^{\infty} \frac{\beta_j(x)}{x} = \frac{h}{2} + \frac{1}{2} \int_0^x q(s)\, ds \quad \text{and} \quad \sum_{j=0}^{\infty} (-1)^j \frac{\beta_j(x)}{x} = \frac{h}{2}$$

(due to the relations $P_j(1) = 1$ and $P_j(-1) = (-1)^j$). The differences

$$\varepsilon_{1,N}(x) := \left| \sum_{j=0}^{N} \frac{\beta_j(x)}{x} - \left(\frac{h}{2} + \frac{1}{2} \int_0^x q(s)\, ds \right) \right| \quad \text{and} \quad \varepsilon_{2,N}(x) := \left| \sum_{j=0}^{N} (-1)^j \frac{\beta_j(x)}{x} - \frac{h}{2} \right|$$
$$\tag{11.16}$$

measure the accuracy of the approximation of the transmutation kernel and hence the accuracy of the approximate solutions (9.14) and (9.15).

Similarly, the accuracy of the coefficients γ_k and the approximations (9.28) and (9.29) can be estimated using (9.23) and the relations [117]

$$K_{h,1}(x, x) = \frac{1}{4}\left(q(x) + h\int_0^x q(s)\,ds + \frac{1}{2}\left(\int_0^x q(s)\,ds\right)^2\right)$$

and

$$K_{h,1}(x, -x) = \frac{1}{4}\left(q(0) + \int_0^x q(s)\,ds\right).$$

The results of Sect. 9.2 allow us to prove the uniform error bound for all approximate zeros of the characteristic function (at least when the coefficients in the boundary conditions (11.10) and (11.11) are independent of the spectral parameter) obtained by the proposed algorithm and that neither spurious zeros appear, nor zeros are missed. For the proof, we refer to [116, Section 7].

The proposed algorithm is based on the exact analytical representation of the solutions (9.12) and (9.13) and their derivatives (9.26) and (9.27). It can be easily combined with widely used techniques such as interval subdivision and the shooting method [142]. However, in order to illustrate that even applied directly the algorithm provides accurate eigendata we present the numerical experiments from [106], which were performed globally without any interval subdivision.

Example 11.1 Consider the following spectral problem (the first Paine problem, [138], see also [116, Example 7.4])

$$\begin{cases} -y'' + e^x y = \lambda y, & 0 \leq x \leq \pi, \\ y(0) = y(\pi) = 0. \end{cases}$$

The criteria from Remark 11.1 suggested to choose $N = 29$. Figure 11.1 shows the absolute and the relative errors of approximation of the first 100 eigenvalues of the problem. Their uniformity is evident.

Figure 11.2 shows the same data, but for the next 400 eigenvalues of the problem. Their uniformity is astonishing. In fact, our numerical experiments show that this picture can be extended to thousands of eigendata.

It is worth mentioning that the whole computation of 500 eigenvalues took less than a second on a PC equipped with the Intel i7-3770 microprocessor.

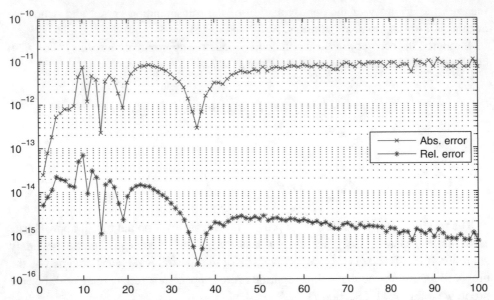

Fig. 11.1 Absolute and relative error of the first 100 eigenvalues from Example 11.1, computed in Matlab, machine precision, $N = 29$

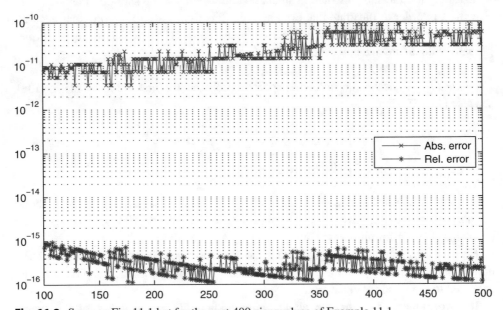

Fig. 11.2 Same as Fig. 11.1 but for the next 400 eigenvalues of Example 11.1

Example 11.2 Consider the following spectral problem from Sect. 11.1 (the second Paine problem, [137], see also [116, Example 7.5]):

$$\begin{cases} -y'' + \frac{1}{(x+0.1)^2}y = \lambda y, & 0 \le x \le \pi, \\ y(0) = y(\pi) = 0. \end{cases}$$

This problem was considered in [116, Example 7.5], where the results were reported for high precision arithmetic only, the reason being the largeness of the coefficients arising in the solution of the approximation problem, which limits the achievable accuracy of the approximation. As a result, saturation occurred starting with $N = 20$ and the eigenvalues were obtained with an error of about 0.001 or worse, see Fig. 11.3.

With the NSBF method (with 84 coefficients β_k) we computed the first 500 eigenvalues with the maximum relative error $5.6 \cdot 10^{-15}$. Figure 11.3 shows the relative errors of the first 100 eigenvalues. The computation time was: 0.69 s for constructing a particular solution and the coefficients β_k, $k \le 100$, and 0.8 s for finding 500 eigenvalues with the help of the secant method.

For more numerical examples and related discussion we refer to [106].

Both methods considered in this chapter have the advantage to offer a characteristic function of the Sturm–Liouville problem in analytical form. The coefficients of the representation are computed using a recurrent integration procedure that is convenient for numerical implementation. The SPPS method is sufficient when a relatively reduced number of eigendata is required to be computed, while the NSBF representation is much more powerful when a large set of the eigendata needs to be found.

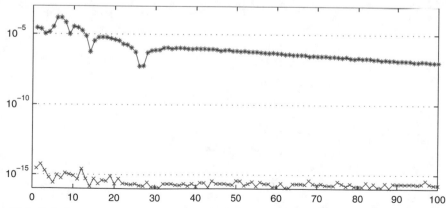

Fig. 11.3 Relative errors of the first 100 eigenvalues from Example 11.2. Upper red line with 'asterisk' marks: the best results we were able to obtain using machine precision and the method from [116], lower blue line with 'x' marks: results obtained by the NSBF method, N = 83

Spectral Problems on Infinite Intervals

12

12.1 Sturm–Liouville Problem on the Half-Line

In this section we consider the Sturm–Liouville problem on the half-line (see Chap. 4) consisting in computation of spectral data of the problem

$$-y'' + q(x)y = \lambda y, \quad x > 0,$$

$$y'(0) - hy(0) = 0, \quad h \in \mathbb{R},$$

where $q(x)$ is a real-valued function satisfying the condition (10.2).

First of all we derive the representations for all spectral data of the problem in terms of the power series with respect to the parameter z defined by (10.8), with the coefficients in the series being the coefficients a_n and d_n, $n = 0, 1, \ldots$ from Chap. 10.

12.1.1 Representations for Spectral Data

To compute the spectral data it is convenient to use the representations (10.9) and (10.27). Indeed, since the eigenvalues (if they exist) are negative numbers, in terms of the parameter z the problem of their computation reduces to finding zeros of the function

$$\Phi(z) := \frac{z-1}{2(z+1)} + \omega(0) + (z+1)\sum_{n=0}^{\infty}(-1)^n z^n d_n(0)$$

$$- h\left(1 + (z+1)\sum_{n=0}^{\infty}(-1)^n z^n a_n(0)\right) \tag{12.1}$$

© The Editor(s) (if applicable) and The Author(s), under exclusive licence to Springer Nature Switzerland AG 2020
V. V. Kravchenko, *Direct and Inverse Sturm-Liouville Problems*,
Frontiers in Mathematics, https://doi.org/10.1007/978-3-030-47849-0_12

when $z \in (-1, 1)$ (when $z \to -1$ one has that $\lambda \to -\infty$, and the equality $z = 1$ corresponds to $\lambda = 0$), $\Phi(z) = \Delta(\rho)$, where $\Delta(\rho)$ was introduced in Chap. 4. The function $\Delta_1(\rho)$ (see Chap. 4) can also be written in terms of z. We have

$$\frac{de(\rho, 0)}{d\rho} = i\,(z+1)^2 \left(\sum_{n=0}^{\infty} (-1)^n\, a_n(0) \left(nz^{n-1} + (n+1)z^n \right) \right),$$

$$\frac{de'(\rho, 0)}{d\rho} = i \left(1 + (z+1)^2 \sum_{n=0}^{\infty} (-1)^n\, d_n(0) \left(nz^{n-1} + (n+1)z^n \right) \right),$$

and thus

$$\Delta_1(\rho) = \frac{1}{2\rho} \frac{d\Delta(\rho)}{d\rho}$$

$$= -\frac{z+1}{z-1} - \frac{(z+1)^3}{z-1} \sum_{n=0}^{\infty} (-1)^n \left(nz^{n-1} + (n+1)z^n \right) (d_n(0) - ha_n(0)). \qquad (12.2)$$

The derivative $V(\lambda)$ of the spectral density function can be computed by the formula

$$V(\lambda) = \frac{i(1-z)}{2\pi(1+z)\,|\Phi(z)|^2}, \qquad (12.3)$$

which is obtained from (4.9). Here z runs counterclockwise along the upper unit semicircle, that is $z = e^{i\theta}$ with $\theta \in (0, \pi)$ ($z = 1$ corresponds to $\lambda = 0$, while $z \to -1$ along the upper unit semicircle corresponds to $\lambda \to \infty$).

Figure 12.1 shows schematically the distribution of the spectrum in terms of the parameter z.

Thus, the spectral data can be computed as follows.

1. For $z \in (-1, 1)$, find the zeros z_k of the function (12.1). Then the eigenvalues are obtained as $\lambda_k = - \left(\frac{z_k - 1}{2(z_k + 1)} \right)^2$.
2. The weight numbers are computed from (4.8) with the aid of (10.9) and (12.2).
3. The function $V(\lambda)$ is computed by (12.3) with $z = e^{i\theta}$ and $\theta \in (0, \pi)$.

12.1.2 Numerical Implementation

We present the results of implementation of the algorithm described in the previous subsection. The algorithm was implemented numerically in Matlab 2017a and reported in [59].

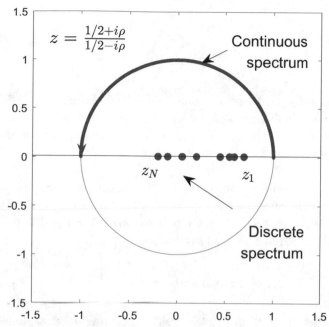

Fig. 12.1 Schematic illustration of the distribution of the spectrum of the Sturm–Liouville problem on the half-line in terms of the parameter z

1. The Jost solution $e\left(\frac{i}{2}, x\right)$ was computed as explained in Sect. 10.5 (and the same for $\eta(x)$). For solving the Cauchy problem the SPPS method was used. All the necessary integrations were performed with the aid of a routine implementing the Newton–Cottes 6-point integration rule.

2. The system of coefficients $\{a_n, a'_n, d_n\}_{n=0,\dots,K-1}$ was computed as described in Sect. 10.5 in combination with (10.29)–(10.34). We emphasize that no numerical differentiation is required at any step.

3. Using the coefficients, all spectral data were computed as described in Sect. 12.1.1.

The control of the accuracy and the choice of an appropriate number K of the coefficients to be computed is performed using (10.6) and (10.35). Indeed, the sufficient smallness of the expressions

$$\varepsilon_{1,K} = \max \left| \sum_{n=0}^{K-1} a_n(x) - \frac{1}{2} \int_x^\infty q(t)dt \right|,$$

$$\varepsilon_{2,K} = \max \left| \sum_{n=0}^{K-1} d_n(x) - \sum_{n=0}^{K-1} \left(n + \frac{1}{2}\right) a_n(x) + \frac{q(x)}{2} \right| \qquad (12.4)$$

and

$$\varepsilon_{3,K} = \max \left| \sum_{n=0}^{K-1} a_n'(x) + \frac{q(x)}{2} \right| \tag{12.5}$$

indicates a sufficiently good approximation of the integral kernels involved.

We notice that all the reported numerical results were obtained within fractions of a second or several seconds, without significant efforts undertaken to optimize the computational time.

Example 12.1 Consider the potential $q(x) = e^{-x}$. With the aid of the asymptotic relations for modified Bessel functions (see, e.g., [136, Sect. 10.30]) it is easy to show that $e(\rho, x) = \Gamma(2\tau + 1)I_{2\tau}(2e^{-\frac{x}{2}})$ when $\rho = i\tau$ and $\tau > 0$, and $e(\rho, x) = \Gamma(-2i\rho + 1)I_{-2i\rho}(2e^{-\frac{x}{2}})$ when $\rho > 0$. Thus, the function $\Delta(\rho)$ has the form

$$\Delta(\rho) = \Gamma(-2i\rho + 1)\left(-\frac{1}{2}\left(I_{-2i\rho-1}(2) + I_{1-2i\rho}(2)\right) - hI_{-2i\rho}(2)\right). \tag{12.6}$$

Substitution of this expression in (4.9) leads to a closed-form expression for $V(\lambda)$. Additionally, when $h = -1$, one eigenvalue exists, $\lambda_1 \approx -0.357735866044089$.

The computation of $e\left(\frac{i}{2}, x\right)$ by the SPPS method did not involve any difficulty (its accuracy was of order 10^{-14}). In this example the method converges extremely fast. In Table 12.1 the absolute error of the computed approximation of the eigenvalue is shown for different numbers of the coefficients participating in the expansion (12.1). $K = 1$ means that $a_0(0)$ and $d_0(0)$ alone participate in (12.1).

In Table 12.2 the maximum absolute error of approximation of $V(\lambda)$, $\lambda > 0$ is presented.

Here we emphasize that while our method (by using (12.3)) computed without any difficulty the function $V(\lambda)$ on the interval $(0, \infty)$ "approximating at infinity" arbitrarily closely, the computation of the exact $V(\lambda)$ presented quite a challenge due to the necessity of computing the modified Bessel functions of complex order ν with $\operatorname{Im}\nu \to -\infty$. Thus, we were able to compare $V(\lambda)$ with $\widetilde{V}(\lambda)$ on a relatively small part of the interval. This can be seen in Fig. 12.2, where the continuous line represents the values of $\widetilde{V}(\lambda)$ computed

Table 12.1 Absolute error of the approximate eigenvalue from Example 12.1

K	$\|\lambda_1 - \widetilde{\lambda}_1\|$
1	0.0013
2	2.9×10^{-5}
3	4.06×10^{-7}
5	1.09×10^{-9}
10	5.37×10^{-13}

Table 12.2 Maximum
absolute error of approximation
of $V(\lambda)$ from Example 12.1

K	$\max |V(\lambda) - \widetilde{V}(\lambda)|$
1	0.0065
2	0.0026
3	4.9×10^{-4}
5	1.09×10^{-4}
10	5.13×10^{-6}
15	6.91×10^{-8}
25	9.75×10^{-10}
35	1.65×10^{-12}
45	3.47×10^{-14}
55	2.49×10^{-14}

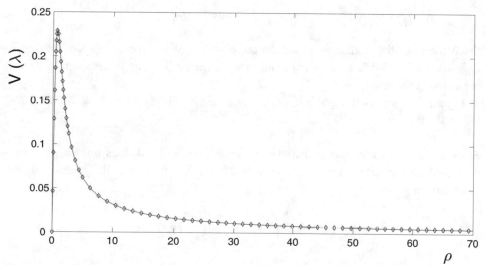

Fig. 12.2 The continuous line represents the values of $\widetilde{V}(\lambda)$ from Example 12.1 computed by (12.3) with $K = 25$ coefficients participating while "open diamond" indicates the values of $V(\lambda)$ calculated by (4.9) and (12.6)

by (12.3) with $K = 25$ coefficients participating while "◇" indicates the values of $V(\lambda)$ calculated by (4.9) and (12.6).

Example 12.2 Consider the potential

$$q(x) = \begin{cases} C, & \text{for } 0 \le x \le 1, \\ 0, & \text{for } x > 1, \end{cases}$$

where C is a positive constant. The Jost solution has the form [9, Example 6.2]

$$e(\rho, x) = \begin{cases} e^{i\rho} \left(\cosh \nu(x-1) + \frac{i\rho}{\nu} \sinh \nu(x-1) \right), & \text{for } 0 \le x \le 1, \\ e^{i\rho x}, & \text{for } x > 1, \end{cases}$$

where $\nu = \sqrt{C - \lambda}$. In particular, for $h = 0$ there exists no eigenvalue, and

$$V(\lambda) = \begin{cases} \dfrac{\rho}{\pi \left| i\rho \cosh \nu - \nu \sinh \nu \right|^2}, & \text{for } 0 < \lambda \le C, \\[3mm] \dfrac{\rho}{\pi \left| i\rho \cos \sqrt{\rho^2 - C} + \sqrt{\rho^2 - C} \sin \sqrt{\rho^2 - C} \right|^2}, & \text{for } \lambda > C. \end{cases} \tag{12.7}$$

Since the potential is discontinuous the convergence of the method is slower than in Example 12.1. In Table 12.3 the maximum absolute error of the approximation of $V(\lambda)$ is presented in the case $C = 3$.

Figure 12.3 shows the graph of $\widetilde{V}(\lambda)$ computed by (12.3) with $K = 50$ with "◇" indicating the values of $V(\lambda)$ calculated by (12.7).

Example 12.3 Consider the potential $q(x) = 8/(1 + 2x)^2$. It does not satisfy condition *(10.2)*. Nevertheless, we tested the method in this example as well. The Jost solution has the form [68]

$$e(\rho, x) = \left(1 - \frac{2}{i\rho(1 + 2x)} \right) e^{i\rho x};$$

for $h = 0$ there exists no eigenvalue, and

$$V(\lambda) = \frac{\rho^3}{\pi \left(\rho^4 - 4\rho^2 + 16 \right)}. \tag{12.8}$$

Figure 12.4 shows the function $V(\lambda)$ computed with $K = 100$, with the symbol "◇" indicating the corresponding exact values. While in the first two examples the truncation

Table 12.3 Maximum absolute error of approximation of $V(\lambda)$ from Example 12.2

K	$\max \left| V(\lambda) - \widetilde{V}(\lambda) \right|$
5	0.085
10	0.049
15	0.027
50	0.006

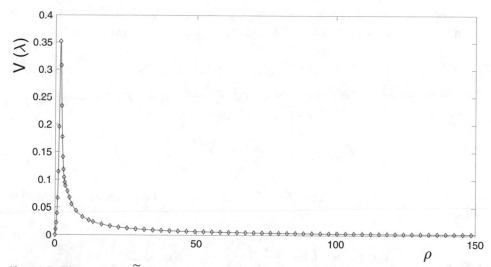

Fig. 12.3 The graph of $\widetilde{V}(\lambda)$ from Example 12.2 computed by (12.3) for $K = 50$ with "open diamond" indicating the values of $V(\lambda)$ calculated by (12.7)

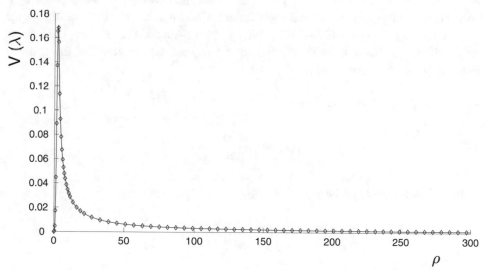

Fig. 12.4 The function $V(\lambda)$ from Example 12.3, computed with $K = 100$. The symbol "open diamond" indicates the corresponding exact values computed by (12.8)

interval need not to be too large, in this example, due to a slow decay of the potential, the truncation interval was chosen equal to 150.

Example 12.4 Consider the potential $q(x) = x/(x^2 + 1)^\mu$ with $\mu > 3/2$. Although the Jost solution is not known in a closed form, in order to check the validity of the numerical results one can estimate the magnitude of $\varepsilon_{j,K}$ from (12.4), (12.5) as well as use the asymptotic relation for the function $V(\lambda)$ from [172, p. 147] according to which

$$V(\lambda) = \frac{1}{\pi \rho} \left(1 + \frac{1}{\rho} \int_0^\infty q(t) \sin 2\rho t \, dt + O\left(\frac{1}{\rho^2}\right) \right), \quad \rho > 0, \quad \rho \to +\infty.$$

With the aid of formula 2.5.9(11) from [141] we find that

$$\int_0^\infty q(t) \sin 2\rho t \, dt = \frac{\sqrt{\pi}}{\Gamma(\mu)} \left(\frac{1}{\rho}\right)^{1/2-\mu} K_{3/2-\mu}(2\rho),$$

and hence we can compare the function $V(\lambda)$ computed to its asymptotic approximation defined as

$$V_a(\lambda) = \frac{1}{\pi \rho} \left(1 + \frac{\sqrt{\pi}}{\Gamma(\mu)} \left(\frac{1}{\rho}\right)^{-1/2-\mu} K_{3/2-\mu}(2\rho) \right).$$

We chose $\mu = \pi$ and $h = -1$. Figure 12.5 shows the convergence of $\varepsilon_{2,K}$ and $\varepsilon_{3,K}$ from (12.4) and (12.5) ($\varepsilon_{1,K}$ converges faster). Choosing $K = 80$, one eigenvalue is computed $\lambda_1 \approx -0.8497040479$ with $\alpha_1 \approx 1.9806595182$. Figure 12.6 shows the fulfillment of the asymptotic relation for $V(\lambda)$, namely, the graph of the quotient $V(\lambda)/V_a(\lambda)$ tends to one when $\rho \to +\infty$.

Example 12.5 Consider the potential

$$q(x) = \frac{1200e^{10x}}{\left(6e^{10x} - 1\right)^2}.$$

From [7, Example 6.1] we have

$$e(\rho, x) = e^{i\rho x} \left(1 + \frac{10i}{\rho + 5i} \frac{1}{6e^{10x} - 1} \right),$$

and hence for $h = 0$ we obtain that

$$\Delta(\rho) = \frac{i\rho(\rho + 7i) - 24i}{\rho + 5i}.$$

It is easy to see that there are no eigenvalues. Figure 12.7 shows the convergence of $\varepsilon_{2,K}$ and $\varepsilon_{3,K}$ from (12.4) and (12.5), together with the magnitude $\max_\lambda |V(\lambda) - \widetilde{V}(\lambda)|$ for $K = 5, \ldots, 175$.

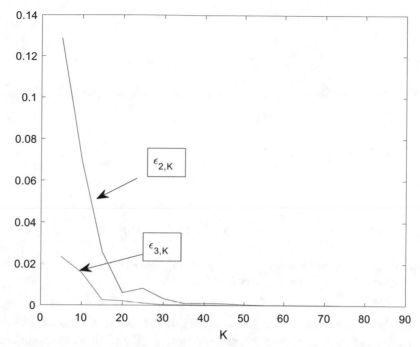

Fig. 12.5 Example 12.4: the convergence of $\varepsilon_{2,K}$ and $\varepsilon_{3,K}$ from (12.4) and (12.5) for $K = 5, \ldots, 85$

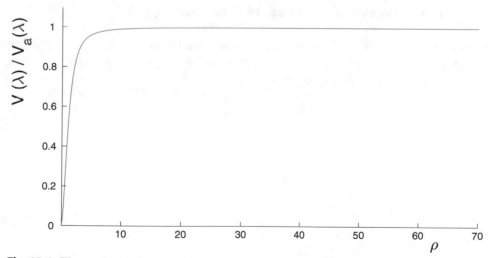

Fig. 12.6 The quotient $V(\lambda)/V_a(\lambda)$ for Example 12.4 tends to one when $\rho \to +\infty$. The function $V(\lambda)$ was computed with $K = 80$

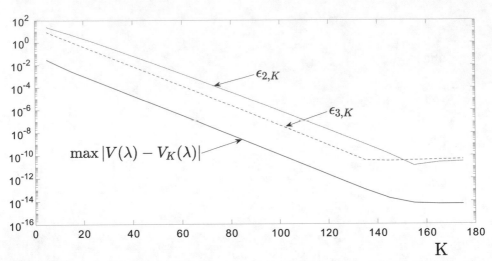

Fig. 12.7 Example 12.5: the convergence of $\varepsilon_{2,K}$ and $\varepsilon_{3,K}$ from (12.4), (12.5) is shown together with the magnitude $\max_\lambda |V(\lambda) - \widetilde{V}(\lambda)|$ for $K = 5, \ldots, 175$

As we show (especially in the last example), if necessary, a large number of the coefficients a_n and d_n can be computed, thus guaranteeing the accuracy of approximation.

All the numerical tests indicate that the presented method allows to compute both the discrete and the continuous spectral data quickly and with a remarkable accuracy. We dare to affirm that no other existing method is able to produce similar results as to the accuracy and easiness of obtaining the spectral data of the problem.

Remark 12.1 The method described in this section can be applied as well to the direct quantum scattering problem from Chap. 5. The scattering data (5.4) are computed then with the aid of formulas from Chap. 5.

Part IV

Solution of Inverse Sturm-Liouville Problems

In this part of the book a unified approach for solving different inverse spectral problems is presented. In each case the corresponding integral transmutation kernel is used in the form of a corresponding functional series representation derived in Part II. For the inverse problems on a finite or semi-infinite interval the integral kernels of the transmutation operators with conditions in the origin from Sect. 8.2 together with their Fourier-Legendre series representations from Sect. 9.2 are used, while for the inverse scattering problem on the line the integral kernel of the transmutation operator with a condition at infinity is used in the form of the Fourier-Laguerre series representation from Sect. 10.1.

The crucial observation is that the potential can be recovered from the very first coefficient of the corresponding functional series representation of the transmutation kernel. In order to compute the first coefficient we derive a system of linear algebraic equations directly from the corresponding Gelfand-Levitan type equation.

This approach for solving a variety of the inverse spectral problems is direct, simple and offers a powerful method for accurate solution of the associated computationally challenging problems. While for the classical inverse Sturm-Liouville problem on a finite interval quite a few numerical methods have been developed (see references in Sect. 3.2), and hence the method proposed here (Chap. 13) can be regarded as just "one more of the pack", albeit with its clear advantages, the situation with the inverse spectral problems on infinite intervals is different. Relatively few modes of numerical solution have been reported. We refer to [8, 12, 68, 155, 165]. The approach presented here leads to a simple and viable numerical technique. The results of the numerical experiments presented below were obtained without much programmer effort for optimizing the computation time, accuracy, etc. Whenever it was possible, standard Matlab routines were employed. Thus, the numerical solution of the inverse spectral problems with the aid of the approach presented here is a simple task which can be considered as an exercise appropriate for undergraduate students.

The Inverse Sturm–Liouville Problem on a Finite Interval

13

13.1 The Gelfand-Levitan Equation

In Sect. 3.2 we introduced the classical inverse Sturm–Liouville problem on a finite interval which is formulated as follows. Given two sequences of real numbers $\{\lambda_n\}_{n=0}^{\infty}$ and $\{\alpha_n\}_{n=0}^{\infty}$ such that $\lambda_n < \lambda_m$ for $n < m$, $\alpha_n > 0$, and the relations (3.5) are valid. Find the real-valued potential $q(x)$ and the real numbers h and H, such that $\{\lambda_n\}_{n=0}^{\infty}$ is the spectrum of the Sturm–Liouville problem

$$-y'' + q(x)y = \lambda y, \quad 0 < x < \pi, \tag{13.1}$$

$$y'(0) - hy(0) = y'(\pi) + Hy(\pi) = 0, \tag{13.2}$$

and α_n, $n = 0, 1, \ldots$ are the corresponding norming constants.

In this chapter we present a method of solution of this problem introduced in [103]. We notice that in [103] an excessive restriction on the potential q (that $q \in C^1[0, \pi]$) was imposed due to the method used to prove the main theorem of this chapter. The method of proof in [103] used some classical results from approximation theory about the rate of convergence of Fourier–Legendre series. In [58] another idea of proof was developed which we reproduce here and which allows us to give a proof for any $q \in L_2(0, \pi)$.

Let $G(x, t)$ be the integral kernel of the transmutation operator T_c (see Sect. 8.2), which transforms the function $\cos \rho x$ into the solution $\varphi(\rho, x)$, of the Cauchy problem (3.3) and (3.4), according to the formula

$$\varphi(\rho, x) = \cos \rho x + \int_0^x G(x, t) \cos \rho t \, dt,$$

© The Editor(s) (if applicable) and The Author(s), under exclusive licence to Springer Nature Switzerland AG 2020
V. V. Kravchenko, *Direct and Inverse Sturm-Liouville Problems*,
Frontiers in Mathematics, https://doi.org/10.1007/978-3-030-47849-0_13

for all $\rho \in \mathbb{C}$.

Let

$$F(x,t) = \sum_{n=0}^{\infty} \left(\frac{\cos \rho_n x \cos \rho_n t}{\alpha_n} - \frac{\cos nx \cos nt}{\alpha_n^0} \right), \qquad (13.3)$$

where

$$\alpha_n^0 = \begin{cases} \pi/2, & \text{if } n > 0, \\ \pi, & \text{if } n = 0. \end{cases}$$

The function $F(x,t)$ is constructed from the given data of the inverse Sturm–Liouville problem. The functions G and F are related by the *Gelfand-Levitan equation* [69,127,172]

$$G(x,t) + F(x,t) + \int_0^x F(t,s)G(x,s)\,ds = 0, \quad 0 < t < x. \qquad (13.4)$$

Remark 13.1 The Gelfand-Levitan equation can be written in the form

$$G(x,t) = -T_c[F](x,t)$$

and hence the function $-F(x,t)$ is the preimage of the function $G(x,t)$ under the transmutation operator T_c. This observation was used in [90] to derive series representations for the transmutation kernel $G(x,t)$ and for the kernel of T_c^{-1} in terms of eigenfunctions of the Sturm–Liouville problem.

Equation (13.4), derived in the famous paper [74], reduces the inverse Sturm–Liouville problem to a Fredholm integral equation of the second kind. Indeed, for each fixed x, (13.4) is a Fredholm equation of the second kind, and when it is solved for the function $G(x,t)$, the potential q can be recovered by the formula

$$q(x) = 2\frac{d}{dx}G(x,x),$$

which is obtained by differentiation of the first equality in (8.7). This apparently simple way of solving the inverse Sturm–Liouville problem has not lead to an efficient numerical algorithm (see the discussion in [130]). The reason lies in the particularities of the convergence of the series in (13.3). In spite of the fact that the function $F(x,t)$ itself is continuous and $\frac{d}{dx}F(x,x) \in L_2(0,\pi)$ [172, p. 43], the series in (13.3) (in the case when the parameter ω defined by (3.6) is not equal to zero) has a jump discontinuity at $x = t = \pi$ (see a detailed explanation in [90]).

By Corollary 9.1,

$$G(x, t) = \sum_{n=0}^{\infty} \frac{g_n(x)}{x} P_{2n}\left(\frac{t}{x}\right), \quad 0 < t \le x, \tag{13.5}$$

where

$$g_0(x) = \varphi(0, x) - 1. \tag{13.6}$$

Thus, if the coefficient $g_0(x)$ is known the potential q can be computed using (13.6):

$$q = \frac{g_0''}{g_0 + 1}. \tag{13.7}$$

The number h is obtained from the equality

$$h = g_0'(0). \tag{13.8}$$

The number H is determined from the formula (3.6) or directly by the formula $H = -\varphi'(\rho_n, \pi)/\varphi(\rho_n, \pi)$, where $\varphi(\rho, x)$ can be calculated by the formula (9.12),

$$\varphi(\rho, x) = \cos \rho x + \sum_{n-0}^{\infty} (-1)^n g_n(x) \, j_{2n}(\rho x).$$

We emphasize that our goal is not to solve the Gelfand-Levitan equation, but rather to use it to obtain a system of linear algebraic equations for the coefficients $g_n(x)$, of which only the very first coefficient $g_0(x)$ is of interest here. Then $q(x)$ is recovered with the aid of (13.7) and the constants h and H are computed as explained above. In the next section we provide a linear algebraic system for computing the coefficients $g_n(x)$.

13.2   A System for the Coefficients g_n

Theorem 13.1 *Let $q \in L_2(0, \pi)$. The coefficients g_n in the expansion (13.5) satisfy the infinite system of linear algebraic equations*

$$\frac{g_k(x)}{4k + 1} + \sum_{m=0}^{\infty} g_m(x) A_{km}(x) = b_k(x) \quad \text{for } k = 0, 1, \ldots, \tag{13.9}$$

where the series

$$A_{km}(x) := (-1)^{m+k} x \sum_{n=0}^{\infty} \left(\frac{j_{2m}(\rho_n x) \, j_{2k}(\rho_n x)}{\alpha_n} - \frac{j_{2m}(nx) \, j_{2k}(nx)}{\alpha_n^0} \right)$$

converge on $[0, \pi]$ to the functions

$$\frac{1}{x} \int_0^x \int_0^x F(t, s) P_{2k}\left(\frac{s}{x}\right) ds \, P_{2m}\left(\frac{t}{x}\right) dt,$$

and the series

$$b_k(x) := (-1)^k x \sum_{n=0}^{\infty} \left(\frac{\cos nx \, j_{2k}(nx)}{\alpha_n^0} - \frac{\cos \rho_n x \, j_{2k}(\rho_n x)}{\alpha_n} \right)$$

converge on $[0, \pi]$ to the functions

$$- \int_0^x F(x, t) P_{2k}\left(\frac{t}{x}\right) dt.$$

For all $k = 0, 1, \ldots$, the series $\sum_{m=0}^{\infty} g_m(x) A_{km}(x)$ converges pointwise on $[0, \pi]$.

Remark 13.2 When $q \in C^1[0, \pi]$, the uniform convergence on $[0, \pi]$ of the series involved can be proved (see [103]).

Proof Substitution of (13.5) into (13.4) gives us the equality

$$\sum_{m=0}^{\infty} \frac{g_m(x)}{x} \left(P_{2m}\left(\frac{t}{x}\right) + \int_0^x F(t, s) P_{2m}\left(\frac{s}{x}\right) ds \right) = -F(x, t). \qquad (13.10)$$

Here we used the fact that fixed x the series in (13.5) converges at least with respect to the norm of $L_2(0, x)$. Consider the integral

$$f_m(x, t) := \int_0^x F(t, s) P_{2m}\left(\frac{s}{x}\right) ds$$

$$= \int_0^x \left(\sum_{n=0}^{\infty} \frac{\cos \rho_n t \cos \rho_n s}{\alpha_n} - \frac{\cos nt \cos ns}{\alpha_n^0} \right) P_{2m}\left(\frac{s}{x}\right) ds. \qquad (13.11)$$

Since for any absolutely continuous function f on $[0, \pi]$

$$\lim_{N \to \infty} \int_0^\pi \left(\sum_{n=0}^{N} \frac{\cos \rho_n t \cos \rho_n s}{\alpha_n} - \frac{\cos nt \cos ns}{\alpha_n^0} \right) f(s) ds = \int_0^\pi F(t, s) f(s) ds$$

uniformly with respect to $t \in [0, \pi]$ (see [69, p. 37], [172, p. 44]), formula (13.11) yields

$$f_m(x, t) = \sum_{n=0}^{\infty} \int_0^x \left(\frac{\cos \rho_n t \cos \rho_n s}{\alpha_n} - \frac{\cos nt \cos ns}{\alpha_n^0} \right) P_{2m} \left(\frac{s}{x} \right) ds$$

$$= \sum_{n=0}^{\infty} \left(\frac{\cos \rho_n t}{\alpha_n} \int_0^x \cos \rho_n s \, P_{2m} \left(\frac{s}{x} \right) ds - \frac{\cos nt}{\alpha_n^0} \int_0^x \cos ns \, P_{2m} \left(\frac{s}{x} \right) ds \right)$$

$$= (-1)^m x \sum_{n=0}^{\infty} \left(\frac{\cos \rho_n t}{\alpha_n} j_{2m} (\rho_n x) - \frac{\cos nt}{\alpha_n^0} j_{2m} (nx) \right),$$

where for the evaluation of the integrals we used formula 2.17.7 from [141, p. 433]:

$$(-1)^m x j_{2m}(\rho x) = \int_0^x \cos \rho s \, P_{2m} \left(\frac{s}{x} \right) ds. \tag{13.12}$$

Note that $b_k(x) = -f_k(x, x)$.
   Thus, Eq. (13.10) takes the form

$$\sum_{m=0}^{\infty} \frac{g_m(x)}{x} \left(P_{2m} \left(\frac{t}{x} \right) + f_m(x, t) \right) = -F(x, t). \tag{13.13}$$

For all $x > 0$ and $0 < t \le x$ the series

$$\sum_{m=0}^{\infty} \frac{g_m(x)}{x} f_m(x, t)$$

converges, since it is the scalar product of the functions $G(x, \cdot)$ and $F(\cdot, t)$ in the space $L_2(0, x)$. Indeed, the system of Legendre polynomials $\left\{ P_{2m} \left(\frac{t}{x} \right) \right\}_{m=0}^{\infty}$ is orthogonal and complete in $L_2(0, x)$. Consequently,

$$F(s, t) = \sum_{n=0}^{\infty} \left\langle F(\cdot, t), P_{2n} \left(\frac{\cdot}{x} \right) \right\rangle \frac{P_{2n} \left(\frac{s}{x} \right)}{\left\| P_{2n} \left(\frac{\cdot}{x} \right) \right\|_{L_2(0,x)}^2}$$

$$= \frac{1}{x} \sum_{n=0}^{\infty} (4n + 1) f_n(x, t) P_{2n} \left(\frac{s}{x} \right).$$

Combining (13.5) with this expression for $F(s, t)$ we obtain that

$$\langle G(x, \cdot), F(\cdot, t) \rangle_{L_2(0,x)} = \sum_{n=0}^{\infty} \left\langle G(x, \cdot), \frac{P_{2n}\left(\frac{\cdot}{x}\right)}{\left\| P_{2n}\left(\frac{\cdot}{x}\right) \right\|} \right\rangle \left\langle F(\cdot, t), \frac{P_{2n}\left(\frac{\cdot}{x}\right)}{\left\| P_{2n}\left(\frac{\cdot}{x}\right) \right\|} \right\rangle$$

$$- \sum_{n=0}^{\infty} \frac{g_n(x)}{x} f_n(x, t),$$

where for the first equality the general Parseval identity [6, p. 16] was used.

Now, returning to Eq. (13.13), we multiply it by $P_{2k}\left(\frac{t}{x}\right)$ and integrate with respect to t from 0 to x. We obtain the equality

$$\frac{g_k(x)}{4k+1} + \sum_{m=0}^{\infty} (-1)^m g_m(x) \times$$

$$\times \sum_{n=0}^{\infty} \left(\frac{j_{2m}(\rho_n x)}{\alpha_n} \int_0^x \cos \rho_n t\, P_{2k}\left(\frac{t}{x}\right) dt - \frac{j_{2m}(nx)}{\alpha_n^0} \int_0^x \cos nt\, P_{2k}\left(\frac{t}{x}\right) dt \right)$$

$$= \sum_{n=0}^{\infty} \left(\frac{\cos nx}{\alpha_n^0} \int_0^x \cos nt\, P_{2k}\left(\frac{t}{x}\right) dt - \frac{\cos \rho_n x}{\alpha_n} \int_0^x \cos \rho_n t\, P_{2k}\left(\frac{t}{x}\right) dt \right).$$

Using again (13.12), we arrive at the equality (13.9), where for all $x > 0$ the series $\sum_{m=0}^{\infty} g_m(x) A_{km}(x)$ converges again due to the general Parseval identity, because it is the scalar product of the functions $G(x, \cdot)$ and $f_k(x, \cdot)$ in $L_2(0, x)$.   ∎

Next, to approximate a number of the coefficients g_k we obtain the linear algebraic system of equations

$$\begin{pmatrix} \widehat{A}_{00} & \widehat{A}_{01} & \dots & \widehat{A}_{0N} \\ \widehat{A}_{10} & \widehat{A}_{11} & \dots & \widehat{A}_{1N} \\ & & \dots & \\ \widehat{A}_{N0} & \widehat{A}_{N1} & \dots & \widehat{A}_{NN} \end{pmatrix} \begin{pmatrix} g_0 \\ g_1 \\ \dots \\ g_N \end{pmatrix} = \begin{pmatrix} \widehat{b}_0 \\ \widehat{b}_1 \\ \dots \\ \widehat{b}_N \end{pmatrix}, \tag{13.14}$$

where

$$\widehat{b}_k(x) = (-1)^k x \sum_{n=0}^{N_s} \left(\frac{\cos nx\, j_{2k}(nx)}{\alpha_n^0} - \frac{\cos \rho_n x\, j_{2k}(\rho_n x)}{\alpha_n} \right), \tag{13.15}$$

$$\widehat{A}_{km}(x) = (-1)^{m+k} x \sum_{n=0}^{N_s} \left(\frac{j_{2m}(\rho_n x)\, j_{2k}(\rho_n x)}{\alpha_n} - \frac{j_{2m}(nx)\, j_{2k}(nx)}{\alpha_n^0} \right) + \frac{\delta_{km}}{4k+1}, \tag{13.16}$$

and δ_{km} is the Kronecker symbol. Here N_s denotes the number of given spectral data $\{\lambda_n, \alpha_n\}_{n=0}^{N_s}$.

Remark 13.3 Thanks to the fact that the kernel $F(x, t)$ belongs to $L_2((0, x) \times (0, x))$, the theory developed in [84, Ch. XIV, Sec. 3, Th. 1] regarding the application of the reduction method to infinite systems of linear algebraic equations is fully applicable to the system (13.9).

13.3   The Method for Recovering the Potential

Given a finite set of spectral data $\{\lambda_n, \alpha_n\}_{n=0}^{N_s}$, the following direct method for recovering the potential q and the numbers h and H is proposed.

1. Compute $\widehat{b}_k$ for $k = 0, 1, \ldots, N$ using (13.15) and $\widehat{A}_{km}$ for $k, m = 0, 1, \ldots, N$ using (13.16).
2. For a set of points $\{x_l\}$ from $(0, \pi]$, solve the system (13.14) and obtain $g_0(x)$.
3. Compute q from (13.7).

All the reported calculations were performed in Matlab 2017a.

Example 13.1 Consider the test problem (from [80]) with $q(x) = \sin 2x$. The "exact spectral data" here and below were computed using the *NSBF* method from Sect. 11.2. The first numerical test was performed with $N_s = 200$ and $N = 6$. Figure 13.1 shows the relative error of the solution $\varphi(0, x) = g_0(x) + 1$ computed in 201 points x_i distributed uniformly in $(0, \pi]$. While in the interior points the error is of order 10^{-5}, it deteriorates in the vicinity of the right endpoint, which is due to the jump discontinuity of the series in (13.3) at $x = t = \pi$.

Next, to obtain an approximation of $q(x)$, the function $\varphi(0, x)$ was approximated by a spline of sixth order and differentiated twice. Figure 13.2 shows the recovered potential. Again in the interior of the interval the approximation is satisfactory (of order 10^{-5}) deteriorating at the end point.

The knowledge of the constant ω from (3.6) which can be approximated using (3.5) is required by several available methods (see, e.g., [80]) and can be helpful in reducing the number of known spectral data. In the next numerical test to 11 "exact spectral data" ($N_s = 10$) we added 190 "asymptotic spectral data" defined by $\rho_n = n + \frac{\omega}{\pi n}$ and $\alpha_n = \frac{\pi}{2}$, $n = 11, \ldots 200$ and applied the same algorithm. Figure 13.3 displays the relative error of the approximation of $\varphi(0, x)$ and Fig. 13.4 the corresponding recovered potential.

The accuracy is still improving with $N_s = 10$ and additional 1990 "asymptotic spectral data", see Figs. 13.5 and 13.6.

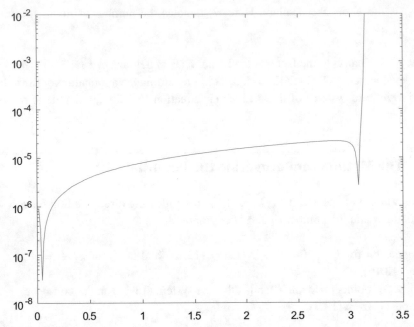

Fig. 13.1 The relative error of the solution $\varphi(0, x) = a_0(x) + 1$ (Example 13.1) computed with $N_s = 200$ and $N = 6$

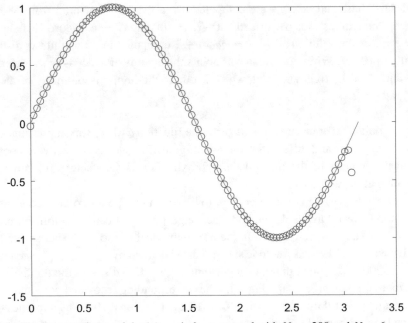

Fig. 13.2 The recovered potential $q(x) = \sin 2x$ computed with $N_s = 200$ and $N = 6$

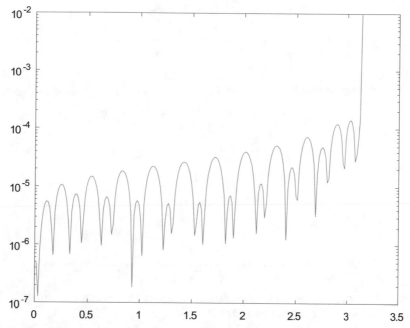

Fig. 13.3 The relative error of the solution $\varphi(0, x)$ (Example 13.1) computed with $N_s = 10$ and additional 190 "asymptotic spectral data", $N = 6$

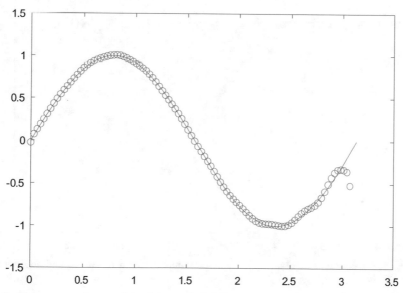

Fig. 13.4 The recovered potential $q(x) = \sin 2x$ computed with $N_s = 10$ and additional 190 "asymptotic spectral data", $N = 6$

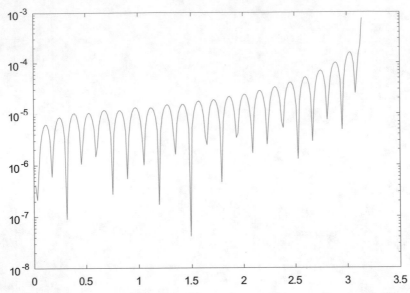

Fig. 13.5 The relative error of the solution $\varphi(0, x)$ (Example 13.1) computed with $N_S = 10$ and additional 1990 "asymptotic spectral data", $N = 6$

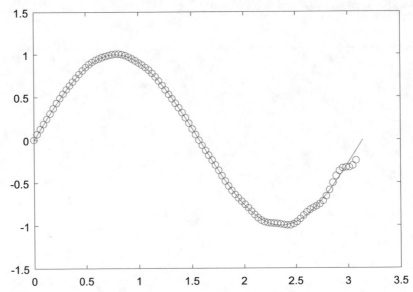

Fig. 13.6 The recovered potential $q(x) = \sin 2x$ computed with $N_S = 10$ and additional 1990 "asymptotic spectral data", $N = 6$

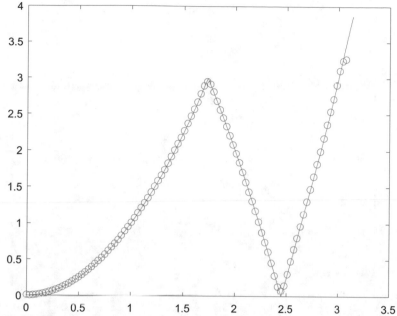

Fig. 13.7 The recovered potential $q(x) = \left|3 - \left|x^2 - 3\right|\right|$ computed with $N_s = 200$ and $N = 6$

Example 13.2 Consider the test problem from [83] with $q(x) = \left|3 - \left|x^2 - 3\right|\right|$. Figure 13.7 displays the recovered potential, computed with $N_s = 200$ and $N = 6$. Figure 13.8 displays the recovered potential, computed with $N_s = 10$, and 1990 "asymptotic spectral data" (see the explanation above) and $N = 6$.

The results demonstrate the satisfactory accuracy of the method as compared to other available numerical approaches, which together with its simplicity makes it worth of further study and application.

Let us notice that the knowledge of the parameter ω can be used for a better recovery of the potential q differently than it was proposed above. Namely, in [90] the series for the function $F(x, t)$ was modified so that it becomes convergent absolutely and uniformly. This can be done if ω is known. Then one can obtain a system analogous to (13.9) for the coefficients g_n starting from that modified series representation of $F(x, t)$ instead of (13.3).

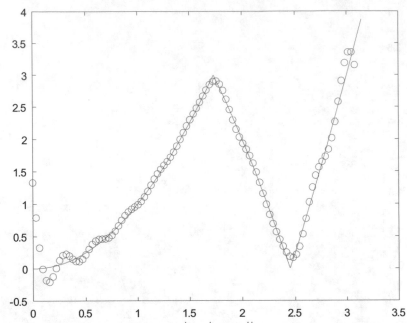

Fig. 13.8 The recovered potential $q(x) = \left|3 - \left|x^2 - 3\right|\right|$ computed with $N = 6$, $N_s = 10$ and 1990 "asymptotic spectral data"

13.4   A Modification of the Method

The numerical experiments from the previous section show that the accuracy of the recovered potential deteriorates at the point $x = \pi$, which is the right end of the interval. This is due to the fact, mentioned in Sect. 13.1, that the series in (13.3) (in the case when the parameter ω defined by (3.6) is not equal to zero) has a jump discontinuity at $x = t = \pi$. To avoid such a deterioration of the accuracy, a simple modification of the method can be recommended. The author thanks Víctor A. Vicente-Benitez for the discussion of this possibility.

The idea of the modification consists in a preparatory integration of the Gelfand-Levitan equation (13.4). Integrating it with respect to the variable t we obtain

$$\int_0^t G(x, \tau)d\tau + \widetilde{F}(x, t) + \int_0^x G(x, s)\widetilde{F}(t, s)\, ds = 0, \quad 0 < t < x \qquad (13.17)$$

where

$$\widetilde{F}(x, t) := \int_0^t F(x, \tau)d\tau.$$

For simplicity, suppose that $\lambda_n \neq 0$ for any $n = 0, 1, \ldots$. Then it is easy to see that

$$\widetilde{F}(x, t) = \frac{\cos \rho_0 x \sin \rho_0 t}{\alpha_0 \rho_0} - \frac{t}{\alpha_0^0} + \sum_{n=1}^{\infty} \left(\frac{\cos \rho_n x \sin \rho_n t}{\alpha_n \rho_n} - \frac{\cos nx \sin nt}{\alpha_n^0 n} \right). \qquad (13.18)$$

The case when $\lambda_n = 0$ for some n requires an obvious change in (13.18), but all subsequent steps remain the same. This series already has no jump discontinuity, which eventually allows us to improve the accuracy of the method at $x = \pi$.

Now the idea is to substitute expressions (13.18) and (13.5) into (13.17). Consider first,

$$\int_0^x P_{2m}\left(\frac{s}{x}\right) \widetilde{F}(t, s)\, ds = \frac{\sin \rho_0 t}{\alpha_0 \rho_0} \int_0^x P_{2m}\left(\frac{s}{x}\right) \cos \rho_0 s\, ds - \frac{t}{\alpha_0^0} \int_0^x P_{2m}\left(\frac{s}{x}\right) ds$$

$$+ \sum_{n=1}^{\infty} \left(\frac{\sin \rho_n t}{\alpha_n \rho_n} \int_0^x P_{2m}\left(\frac{s}{x}\right) \cos \rho_n s\, ds - \frac{\sin nt}{\alpha_n^0 n} \int_0^x P_{2m}\left(\frac{s}{x}\right) \cos ns\, ds \right)$$

$$= -\frac{xt}{\pi} \delta_{0m} + (-1)^m x \left(\frac{\sin \rho_0 t}{\alpha_0 \rho_0} j_{2m}(\rho_0 x) + \sum_{n=1}^{\infty} \left(\frac{\sin \rho_n t}{\alpha_n \rho_n} j_{2m}(\rho_n x) - \frac{2 \sin nt}{\pi n} j_{2m}(nx) \right) \right),$$

where δ_{0m} is the Kronecker symbol.

For the first term in (13.17) we have

$$\int_0^t G(x, \tau)\, d\tau = \sum_{n=0}^{\infty} \frac{g_n(x)}{x} \int_0^t P_{2n}\left(\frac{\tau}{x}\right) d\tau$$

$$= \frac{g_0(x) t}{x} + \sum_{n=1}^{\infty} \frac{g_n(x)}{4n + 1} \left(P_{2n+1}\left(\frac{t}{x}\right) - P_{2n-1}\left(\frac{t}{x}\right) \right).$$

Here we used the identity

$$\int_0^t P_{2n}\left(\frac{\tau}{x}\right) d\tau = \frac{x}{4n + 1} \left(P_{2n+1}\left(\frac{t}{x}\right) - P_{2n-1}\left(\frac{t}{x}\right) \right), \qquad n = 1, 2, \ldots.$$

Thus, the integrated Gelfand-Levitan equation (13.17) takes the form

$$\frac{g_0(x)t}{x} + \sum_{n=1}^{\infty} \frac{g_n(x)}{4n+1}\left(P_{2n+1}\left(\frac{t}{x}\right) - P_{2n-1}\left(\frac{t}{x}\right)\right)$$

$$+ \sum_{m=0}^{\infty} g_m(x)\left[-\frac{t}{\pi}\delta_{0m} + (-1)^m\left(\frac{\sin\rho_0 t}{\alpha_0\rho_0}j_{2m}(\rho_0 x)\right.\right.$$

$$\left.\left. + \sum_{n=1}^{\infty}\left(\frac{\sin\rho_n t}{\alpha_n\rho_n}j_{2m}(\rho_n x) - \frac{2\sin nt}{\pi n}j_{2m}(nx)\right)\right)\right]$$

$$= \frac{t}{\pi} - \frac{\cos\rho_0 x \sin\rho_0 t}{\alpha_0\rho_0} - \sum_{n=1}^{\infty}\left(\frac{\cos\rho_n x \sin\rho_n t}{\alpha_n\rho_n} - \frac{2\cos nx \sin nt}{\pi n}\right).$$

Multiplying this equality by $P_{2k+1}\left(\frac{t}{x}\right)$ and integrating with respect to t from 0 to x, we obtain for $k = 0$ the equation

$$\frac{g_0(x)}{3} - \frac{g_1(x)}{15} +$$

$$\sum_{m=0}^{\infty}(-1)^m g_m(x)\left[\frac{j_{2m}(\rho_0 x)j_1(\rho_0 x)}{\alpha_0\rho_0} - \frac{x\delta_{0m}}{3\pi}\right.$$

$$\left. + \sum_{n=1}^{\infty}\left(\frac{j_{2m}(\rho_n x)j_1(\rho_n x)}{\alpha_n\rho_n} - \frac{2j_{2m}(nx)j_1(nx)}{\pi n}\right)\right]$$

$$= \frac{x}{3\pi} - \frac{\cos\rho_0 x \, j_1(\rho_0 x)}{\alpha_0\rho_0} - \sum_{n=1}^{\infty}\left(\frac{\cos\rho_n x \, j_1(\rho_n x)}{\alpha_n\rho_n} - \frac{2\cos nx \, j_1(nx)}{\pi n}\right),$$

and for $k = 1, 2, \ldots$ the equation

$$\frac{g_k(x)}{(4k+1)(4k+3)} - \frac{g_{k+1}(x)}{(4k+3)(4k+5)} +$$

$$\sum_{m=0}^{\infty}(-1)^{m+k} g_m(x)\left[\frac{j_{2m}(\rho_0 x)j_{2k+1}(\rho_0 x)}{\alpha_0\rho_0}\right.$$

$$\left. + \sum_{n=1}^{\infty}\left(\frac{j_{2m}(\rho_n x)j_{2k+1}(\rho_n x)}{\alpha_n\rho_n} - \frac{2j_{2m}(nx)j_{2k+1}(nx)}{\pi n}\right)\right]$$

$$= (-1)^k\left[-\frac{\cos\rho_0 x \, j_{2k+1}(\rho_0 x)}{\alpha_0\rho_0} - \sum_{n=1}^{\infty}\left(\frac{\cos\rho_n x \, j_{2k+1}(\rho_n x)}{\alpha_n\rho_n} - \frac{2\cos nx \, j_{2k+1}(nx)}{\pi n}\right)\right].$$

Thus, the coefficients $g_k(x)$ satisfy the system of equations

$$\sum_{m=0}^{\infty} g_m(x)C_{km}(x) = d_k(x), \qquad \text{for all } k = 0, 1, \ldots, \tag{13.19}$$

where

$$C_{km}(x) = \begin{cases} c_{k,k}(x) + \frac{1}{(4k+1)(4k+3)}, & \text{if } m = k, \\ c_{k,k+1}(x) - \frac{1}{(4k+3)(4k+5)}, & \text{if } m = k+1, \\ c_{k,m}(x), & \text{otherwise}, \end{cases}$$

$$c_{0,m}(x) = (-1)^m \left[\frac{j_{2m}(\rho_0 x) j_1(\rho_0 x)}{\alpha_0 \rho_0} - \frac{x \delta_{0m}}{3\pi} \right. $$
$$\left. + \sum_{n=1}^{\infty} \left(\frac{j_{2m}(\rho_n x) j_1(\rho_n x)}{\alpha_n \rho_n} - \frac{2 j_{2m}(nx) j_1(nx)}{\pi n} \right) \right],$$

$$c_{k,m}(x) = (-1)^{m+k} \left[\frac{j_{2m}(\rho_0 x) j_{2k+1}(\rho_0 x)}{\alpha_0 \rho_0} \right.$$
$$\left. + \sum_{n=1}^{\infty} \left(\frac{j_{2m}(\rho_n x) j_{2k+1}(\rho_n x)}{\alpha_n \rho_n} - \frac{2 j_{2m}(nx) j_{2k+1}(nx)}{\pi n} \right) \right]$$

and

$$d_k(x) = (-1)^k \left[\frac{x \delta_{0k}}{3\pi} - \frac{\cos \rho_0 x \, j_{2k+1}(\rho_0 x)}{\alpha_0 \rho_0} \right.$$
$$\left. - \sum_{n=1}^{\infty} \left(\frac{\cos \rho_n x \, j_{2k+1}(\rho_n x)}{\alpha_n \rho_n} - \frac{2 \cos nx \, j_{2k+1}(nx)}{\pi n} \right) \right].$$

Example 13.3 Consider the test problem from Example 13.1. Take $N_s = 200$ and $N = 6$. Figure 13.9 shows the recovered potential computed in 21 points distributed uniformly on $[0, \pi]$. Thus, the problem with the deteriorating accuracy at the right endpoint is fixed (cf. Fig. 13.2). The accuracy in the interior of the interval remains of the same order as when the system (13.9) is used.

Similar results are obtained for other test problems.

Example 13.4 Consider the potential from Example 13.2. Under the same conditions as those generating Fig. 13.7, the Figure 13.10 shows the recovered potential, computed with the aid of the system (13.19). Again we observe that the problem with the deteriorating accuracy at π is fixed.

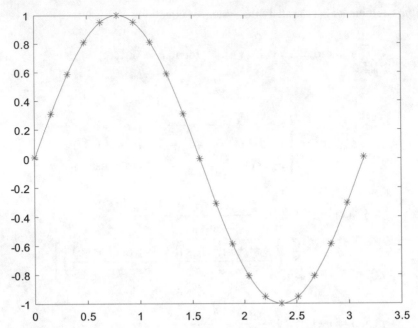

Fig. 13.9 The potential from Example 13.1 computed in 21 points with $N_s = 200$, $N = 6$ by solving the system (13.19) obtained from the integrated Gelfand-Levitan equation

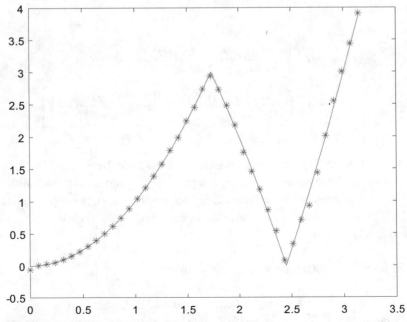

Fig. 13.10 The potential from Example 13.2 computed under the same conditions as 13.7, but with the aid of the system (13.19) obtained from the integrated Gelfand-Levitan equation

Solution of the Inverse Problem on the Half-Line   **14**

14.1   The Gelfand-Levitan Equation

In the present chapter we consider the inverse Sturm-Liouville problem on the half-line introduced in Chap. 4. Thus, we assume that the set of the spectral data

$$\left\{ V(\lambda), \ \lambda > 0; \quad \{\lambda_n, \ \alpha_n\}_{n=\overline{1,N}} \right\}$$

is known, where $V(\lambda)$ is continuous and positive for $\lambda > 0$, the eigenvalues λ_n, if they exist, are negative, and $\alpha_n > 0$. The function $V(\lambda)$ has the asymptotics [69], [172, p. 147]

$$V(\lambda) = \frac{1}{\pi \rho} \left(1 + o\left(\frac{1}{\rho}\right) \right), \quad \rho > 0, \quad \rho \to \infty. \tag{14.1}$$

The method described below was developed in [58].

Similarly to the previous chapter, we will use the integral kernel $G(x, t)$ of the transmutation operator T_c, which satisfies the Gelfand-Levitan equation (see, e.g., [69, 127, 172])

$$G(x, y) + F(x, y) + \int_0^x G(x, t) F(t, y)\, dt = 0, \quad 0 < y \le x < \infty, \tag{14.2}$$

where now the function F contains the spectral data of the inverse Sturm-Liouville problem on the half-line:

$$F(x, y) := \int_{-\infty}^{\infty} \cos \rho x \cos \rho y \, d\widehat{\sigma}(\lambda), \tag{14.3}$$

© The Editor(s) (if applicable) and The Author(s), under exclusive licence to Springer Nature Switzerland AG 2020
V. V. Kravchenko, *Direct and Inverse Sturm-Liouville Problems*,
Frontiers in Mathematics, https://doi.org/10.1007/978-3-030-47849-0_14

where $\widehat{\sigma}(\lambda)$ is defined by

$$\widehat{\sigma}(\lambda) := \begin{cases} \sigma(\lambda) - \sigma_0(\lambda), & \text{if } \lambda \geq 0, \\ \sigma(\lambda), & \text{if } \lambda < 0, \end{cases}$$

and

$$\sigma_0(\lambda) = \frac{2}{\pi}\rho, \quad \lambda \geq 0$$

is the spectral density function of the problem

$$-y'' = \lambda y, \quad 0 < x < \infty, \tag{14.4}$$

$$y(0) = 1, \quad y'(0) = 0. \tag{14.5}$$

Equivalently, [69]

$$d\widehat{\sigma}(\lambda) = \begin{cases} \left(V(\lambda) - \frac{1}{\pi\rho}\right)d\lambda, & \text{if } \lambda > 0 \\ \alpha_k, & \text{if } \lambda = \lambda_k < 0, \quad k = 1, \ldots, N. \end{cases} \tag{14.6}$$

As in the preceding chapter, we use the Fourier-Legendre series representation of $G(x, t)$ from Corollary 9.1,

$$G(x, t) = \sum_{n=0}^{\infty} \frac{g_n(x)}{x} P_{2n}\left(\frac{t}{x}\right), \quad 0 < t \leq x < \infty, \tag{14.7}$$

and the possibility to recover the potential $q(x)$ directly from the first coefficient of the series:

$$q(x) = \frac{g_0''(x)}{g_0(x) + 1} \tag{14.8}$$

as well as the formula

$$h = g_0'(0). \tag{14.9}$$

Thus, the inverse Sturm-Liouville problem reduces to the problem of finding the coefficient g_0. In the next section a system of linear algebraic equations for the coefficients g_n is derived which allows us to solve this problem.

14.2   Linear System of Equations for the Coefficients g_n

Theorem 14.1 *The coefficients $g_n(x)$ from the Fourier-Legendre series representation (14.7) satisfy the system of equations*

$$\frac{g_m(x)}{4m+1} + \sum_{n=0}^{\infty} g_n(x) A_{m,n}(x) = -f_m(x,x), \quad \text{for all } m = 0, 1, \ldots, \tag{14.10}$$

where

$$A_{m,n}(x) := (-1)^{n+m} x \int_{-\infty}^{\infty} j_{2n}(\rho x) j_{2m}(\rho x) \, d\widehat{\sigma}(\lambda), \tag{14.11}$$

$$f_n(x, y) := (-1)^n x \int_{-\infty}^{\infty} j_{2n}(\rho x) \cos \rho y \, d\widehat{\sigma}(\lambda), \quad 0 < y \le x < \infty, \tag{14.12}$$

and j_k stands for the spherical Bessel function of order k.

Proof Substitution of (14.7) into the Gelfand-Levitan equation (14.2) leads to the equality

$$\sum_{n=0}^{\infty} \frac{g_n(x)}{x} \left(P_{2n} \left(\frac{y}{x} \right) + \int_0^x F(t, y) P_{2n} \left(\frac{t}{x} \right) dt \right) = -F(x, y). \tag{14.13}$$

Observe that

$$\int_0^x F(t, y) P_{2n} \left(\frac{t}{x} \right) dt = f_n(x, y). \tag{14.14}$$

Indeed, substitution of (14.3) into this integral leads to the equality

$$\int_0^x F(t, y) P_{2n} \left(\frac{t}{x} \right) dt = \int_0^x \int_{-\infty}^{\infty} \cos \rho t \cos \rho y \, d\widehat{\sigma}(\lambda) P_{2n} \left(\frac{t}{x} \right) dt.$$

Due to (14.1), the integral $\int_{-\infty}^{\infty} \cos \rho t \cos \rho y \, d\widehat{\sigma}(\lambda)$ is absolutely convergent, and hence Fubini's theorem can be applied to change the order of integration. Using the identity [141, formula 2.17.7]

$$(-1)^n x j_{2n}(\rho x) = \int_0^x \cos \rho y P_{2n} \left(\frac{y}{x} \right) dy \tag{14.15}$$

we obtain (14.14), and then (14.13) takes the form

$$\sum_{n=0}^{\infty} \frac{g_n(x)}{x} \left(P_{2n} \left(\frac{y}{x} \right) + f_n(x, y) \right) = -F(x, y). \tag{14.16}$$

For all $x > 0$ and $0 < y \leq x$ the series

$$\sum_{n=0}^{\infty} \frac{g_n(x)}{x} f_n(x, y)$$

converges, since it is the inner product of $G(x, \cdot)$ and $F(\cdot, y)$ in $L_2(0, x)$. Indeed, since the set of the Legendre polynomials $\left\{ P_{2n} \left(\frac{\cdot}{x} \right) \right\}_{n=0}^{\infty}$ is complete and orthogonal in $L_2(0, x)$, together with (14.7) we have that

$$F(t, y) = \sum_{n=0}^{\infty} \left\langle F(\cdot, y), P_{2n} \left(\frac{\cdot}{x} \right) \right\rangle \frac{P_{2n} \left(\frac{t}{x} \right)}{\left\| P_{2n} \left(\frac{\cdot}{x} \right) \right\|_{L_2(0,x)}^2} = \frac{1}{x} \sum_{n=0}^{\infty} (4n + 1) f_n(x, y) P_{2n} \left(\frac{t}{x} \right)$$

Consequently (due to the general Parseval relation [6, p. 16]),

$$\langle G(x, \cdot), F(\cdot, y) \rangle_{L_2(0,x)} = \sum_{n=0}^{\infty} \left\langle G(x, \cdot), \frac{P_{2n} \left(\frac{\cdot}{x} \right)}{\left\| P_{2n} \left(\frac{\cdot}{x} \right) \right\|} \right\rangle \left\langle F(\cdot, y), \frac{P_{2n} \left(\frac{\cdot}{x} \right)}{\left\| P_{2n} \left(\frac{\cdot}{x} \right) \right\|} \right\rangle$$

$$= \sum_{n=0}^{\infty} \frac{g_n(x)}{x} f_n(x, y).$$

Now, multiplying (14.16) by $P_{2m} \left(\frac{y}{x} \right)$ and integrating with respect to y we obtain

$$\frac{g_m(x)}{4m + 1} + \sum_{n=0}^{\infty} (-1)^n g_n(x) \int_{-\infty}^{\infty} j_{2n}(\rho x) \left(\int_0^x P_{2m} \left(\frac{y}{x} \right) \cos \rho y \, dy \right) d\widehat{\sigma}(\lambda)$$

$$= - \int_{-\infty}^{\infty} \cos \rho x \left(\int_0^x P_{2m} \left(\frac{y}{x} \right) \cos \rho y \, dy \right) d\widehat{\sigma}(\lambda).$$

Applying (14.15) we obtain (14.10), where for all $x > 0$ the series $\sum g_n(x) A_{m,n}(x)$ converges, again due to the general Parseval equality, as the inner product of $G(x, \cdot)$ and $f_m(x, \cdot)$ in $L_2(0, x)$. ∎

Remark 14.1 Using (14.6) it is easy to decompose the expressions $A_{m,n}(x)$ and $f_n(x, y)$ into the parts associated with the continuous and the discrete spectra. Namely,

$$A_{m,n}(x) = (-1)^{n+m} x \left(\int_0^\infty j_{2n}(\rho x) j_{2m}(\rho x) \hat{V}(\lambda) \, d\lambda + \sum_{k=1}^N \alpha_k j_{2n}(\rho_k x) j_{2m}(\rho_k x) \right),$$

(14.17)

and

$$f_n(x, y) = (-1)^n x \left(\int_0^\infty j_{2n}(\rho x) \cos \rho y \, \hat{V}(\lambda) \, d\lambda + \sum_{k=1}^N \alpha_k j_{2n}(\rho_k x) \cos \rho_k y \right),$$

(14.18)

where

$$\hat{V}(\lambda) = V(\lambda) - \frac{1}{\pi \rho}.$$

14.3 Numerical Realization and Examples

Theorem 14.1 together with equalities (14.8) and (14.9) lead to a simple and direct numerical algorithm for solving the inverse Sturm-Liouville problem on a half-line. Given the spectral data,

(1) compute $f_m^\Lambda(x, x)$ and $A_{m,n}^\Lambda(x)$ by the formulas

$$f_m^\Lambda(x, x) = (-1)^n x \left(\int_0^\Lambda j_{2m}(\rho x) \cos \rho x \, \hat{V}(\lambda) \, d\lambda + \sum_{k=1}^N \alpha_k j_{2n}(\rho_k x) \cos \rho_k x \right),$$

(14.19)

$$A_{m,n}^\Lambda(x) = (-1)^{n+m} x \left(\int_0^\Lambda j_{2n}(\rho x) j_{2m}(\rho x) \hat{V}(\lambda) \, d\lambda + \sum_{k=1}^N \alpha_k j_{2n}(\rho_k x) j_{2m}(\rho_k x) \right),$$

(14.20)

$m, n = 0, 1, \ldots, M;$

(2) for a set of points $\{x_l\}$ from $(0, b]$, $b > 0$, solve the system

$$\frac{g_m(x)}{4m+1} + \sum_{n=0}^{M} g_n(x) A_{m,n}^{\Lambda}(x) = -f_m^{\Lambda}(x, x), \quad m = 0, 1, \ldots, M \tag{14.21}$$

obtaining $g_0(x)$;

(3) compute $q(x)$, $x \in (0, b)$ and h from (14.8) and (14.9), respectively.

Since the decay of the integrand in (14.19) and (14.20) depends on x, it is convenient to choose Λ inversely proportional to x. We fix a constant $P > 0$ and take $\Lambda(x) = P/x$.

All the reported calculations were performed in Matlab 2017a. For the integration required in step (1) the Matlab routine '*integral*' was used. The differentiation of $g_0(x)$ at step (3) was performed by converting $g_0(x)$ into a spline with the aid of the routine '*spapi*', with a subsequent application of the routine '*fnder*'.

Example 14.1 Consider a test problem with

$$q(x) = \frac{1200e^{10x}}{(6e^{10x} - 1)^2}, \quad x > 0.$$

The Jost solution is given by *(see [7])*

$$e(\rho, x) = e^{i\rho x}\left(\frac{10i}{\rho + 5i} \cdot \frac{1}{6e^{10x} - 1}\right), \quad x > 0.$$

For $h = 0$, the function V has the form

$$V(\lambda) = \frac{(\lambda + 25)\rho}{\pi|\lambda + 7i\rho - 24|^2}, \quad \lambda > 0$$

and the discrete spectrum is empty.

With $P = 150$ and 2001 points $\{x_l\}$ uniformly distributed in the interval $[0, \pi]$, the maximum error of approximation of the potential q by the recovered potential *(denoted by* q_M*)* is reported in Table 14.1.

Example 14.2 Let β be a positive number. Consider *(see [7])*

$$q(x) = \frac{-8\beta^2 e^{2\beta x}}{(e^{2\beta x} + 1)^2}, \quad x > 0.$$

The Jost solution is given by

$$e(\rho, x) = e^{i\rho x}\left(1 - \frac{2i\beta}{\rho + i\beta}\frac{1}{e^{2\beta x} + 1}\right), \quad x > 0.$$

Table 14.1 Maximum absolute error of approximation of the potential q from Example 14.1 by the recovered potential

M	$\max_{x \in (0,\pi)} |q(x) - q_M(x)|$
0	2.017
1	0.109
2	0.015
3	0.0035
4	0.0010
5	3.5068×10^{-5}
7	3.5068×10^{-5}

For $h = 0$, the function V has the form

$$V(\lambda) = \frac{\rho}{\pi(\lambda + \beta^2)}, \quad \lambda > 0.$$

The only eigenvalue is $\lambda_1 = -\beta^2 < 0$ and its corresponding weight number is $\alpha_1 = \beta > 0$. We chose $\beta = \pi$. For 2001 uniformly distributed points on $[0, 1]$ and $P = 150$ already with $M = 0$ (that is, system (14.21) reduces to a single equation) the maximal error of approximation of the potential q by the recovered potential q_0 was 7.962×10^{-6} and did not improve for greater M. This possibly happens due to some special structure of the transmutation kernel in this example.

Example 14.3 Consider the potential $q(x) = 8/(1 + 2x)^2$ with $h = 0$. The function V has the form

$$V(\lambda) = \frac{\lambda^{3/2}}{\pi(\lambda^2 - 4\lambda + 16)}, \quad \lambda \geq 0,$$

and no eigenvalue exists.

In [68, Ex. 3.2] this example was used to test an iterative numerical algorithm developed there. The absolute error of the approximation of the potential on the interval $(0, \pi)$ reported in [68, Ex. 3.2] was 0.013.

With $P = 200$ and 1001 points $\{x_l\}$ uniformly distributed in the segment $[0, \pi]$ the maximal error of approximation of the potential q by the recovered potential (denoted by q_M) rapidly stabilizes at the order 10^{-6} as reported in Table 14.2.

Table 14.2 Maximum absolute error of approximation of the potential q from Example 14.3 by the recovered potential

M	$\max_{x \in (0,\pi)} |q(x) - q_M(x)|$
0	1.723
1	7.56×10^{-6}
2	1.44×10^{-6}
4	1.44×10^{-6}

Solution of the Inverse Quantum Scattering Problem on the Half-Line

15.1 The Gelfand-Levitan Equation

Here we consider the inverse problem formulated in Chap. 5. Given the scattering data (5.4), find the corresponding short-range potential q.

Let $\psi(\rho, x)$ denote a solution of (4.1) satisfying the initial conditions

$$\psi(\rho, 0) = 0, \quad \psi'(\rho, 0) = 1. \tag{15.1}$$

As shown in Sect. 8.2, $\psi(\rho, x)$ admits the Povzner-Levitan representation

$$\psi(\rho, x) = T_s\left[\frac{\sin \rho x}{\rho}\right] = \frac{\sin \rho x}{\rho} + \int_0^x S(x, t)\frac{\sin \rho t}{\rho}dt$$

where

$$S(x, t) = -\frac{1}{\pi}\int_0^\infty \left(\psi(\rho, x) - \frac{\sin \rho x}{\rho}\right)\sin \rho t \, \rho d\rho,$$

$S(x, 0) = 0$, $S(x, x) = \frac{1}{2}\int_0^x q(t)dt$ [51].

The inverse problem can be solved via the Gelfand-Levitan equation (see, e.g., [51, Chapter III])

$$S(x, y) + F(x, y) + \int_0^x S(x, s)F(s, y)ds = 0, \tag{15.2}$$

© The Editor(s) (if applicable) and The Author(s), under exclusive licence to Springer
Nature Switzerland AG 2020
V. V. Kravchenko, *Direct and Inverse Sturm-Liouville Problems*,
Frontiers in Mathematics, https://doi.org/10.1007/978-3-030-47849-0_15

where

$$F(x, y) = \frac{2}{\pi} \int_0^\infty \sin \rho x \sin \rho y \left(\frac{1}{|F(\rho)|^2} - 1 \right) d\rho + \sum_j c_j \sinh(i\rho_j x) \sinh(i\rho_j y)$$

(15.3)

and

$$c_j := \frac{1}{\int_0^\infty \psi^2(\rho_j, x) dx} = -\frac{2\rho_j e'(\rho_j, 0)}{F'(\rho_j)}.$$

Here we assume that $F(\rho)$ is given. It can be obtained from the scattering matrix using the relations (I.4.11) and (II.2.1) from [51], although this represents an additional computational challenge. We mention that another possibility is to solve the inverse quantum scattering problem with the aid of the corresponding Marchenko equation. This approach is somewhat simpler than the approach based on the Gelfand-Levitan equation. The reason is that the input integral kernel is obtained directly from the scattering data. However, since in the next chapter we consider the inverse scattering problem on the line, for which the Marchenko equation is, in fact, the only possibility, here we present the approach based on the Gelfand-Levitan equation.

By Corollary 9.1, the following Fourier–Legendre representation for the kernel $S(x, y)$ is valid:

$$S(x, y) = \sum_{n=0}^\infty \frac{s_n(x)}{x} P_{2n+1}\left(\frac{y}{x}\right), \quad 0 < y \le x < \infty.$$

(15.4)

For each fixed $x > 0$, the series converges in the norm of $L_2(0, x)$. The coefficients s_n can be computed by a recurrent integration procedure, starting with

$$s_0(x) = \frac{3}{2} \left(\frac{\psi(0, x)}{x} - 1 \right).$$

(15.5)

Again we emphasize that for solving the inverse problem knowing the recurrent integration procedure is superfluous: only (15.4) and (15.5) are required. Notice that equality (15.5) indicates that the first coefficient s_0 is sufficient for recovering the potential q. Indeed, since $\psi''(0, x) = q(x)\psi(0, x)$, we have that

$$q(x) = \frac{(xs_0(x))''}{\left(xs_0(x) + \frac{3}{2}x \right)}.$$

(15.6)

15.2   Linear System of Equations for the Coefficients s_n

In this section we derive a system of linear algebraic equations for the coefficients $\{s_n\}_{n=0}^{\infty}$; in fact, we are interested in computing the very first coefficient s_0.

Theorem 15.1 ([85]) *The coefficients of (15.4) satisfy the system of equations*

$$\frac{s_m(x)}{4m+3} + \sum_{n=0}^{\infty} s_n(x) B_{m,n}(x) = -f_m(x,x), \quad \text{for all } m = 0, 1, \ldots, \tag{15.7}$$

where

$$B_{m,n}(x) := (-1)^{n+m} \frac{2x}{\pi} \int_0^{\infty} j_{2n+1}(\rho x) j_{2m+1}(\rho x) \left(\frac{1}{|F(\rho)|^2} - 1 \right) d\rho$$

$$+ (-1)^{n+m} ix \sum_j c_j j_{2n+1}(\rho_j x) j_{2m+1}(\rho_j x), \tag{15.8}$$

$$f_m(x,y) := (-1)^m \frac{2x}{\pi} \int_0^{\infty} j_{2m+1}(\rho x) \sin(\rho y) \left(\frac{1}{|F(\rho)|^2} - 1 \right) d\rho$$

$$+ (-1)^m ix \sum_j c_j j_{2m+1}(\rho_j x) \sinh(i\rho_j y), \tag{15.9}$$

$0 < y \le x < \infty$.

Proof The proof is similar to that from the previous chapter. Substitution of (15.4) into the Gelfand-Levitan equation (15.2) leads to the equality

$$\sum_{n=0}^{\infty} \frac{s_n(x)}{x} \left(P_{2n+1}\left(\frac{y}{x}\right) + \int_0^x F(t,y) P_{2n+1}\left(\frac{t}{x}\right) dt \right) = -F(x,y). \tag{15.10}$$

Note that

$$\int_0^x F(t,y) P_{2n+1}\left(\frac{t}{x}\right) dt = f_n(x,y). \tag{15.11}$$

Indeed, substitution of (15.3) into this integral yields

$$\int_0^x F(t,y) P_{2n+1}\left(\frac{t}{x}\right) dt = \frac{2}{\pi} \int_0^{\infty} \sin(\rho y) \left(\frac{1}{|F(\rho)|^2} - 1 \right) \int_0^x \sin(\rho t) P_{2n+1}\left(\frac{t}{x}\right) dt d\rho$$

$$+ \sum_j c_j \sinh(i\rho_j y) \int_0^x \sinh(i\rho_j t) P_{2n+1}\left(\frac{t}{x}\right) dt,$$

which with the aid of [141, formula 2.17.7.1] gives (15.11). Here (5.2) ensures that the integral $\int_0^\infty \sin(\rho t) \sin(\rho y) \left(\frac{1}{|F(\rho)|^2} - 1 \right) d\rho$ is absolutely convergent, and hence Fubini's theorem can be applied to change the order of integration.

Hence (15.10) takes the form

$$\sum_{n=0}^{\infty} \frac{s_n(x)}{x} \left(P_{2n+1} \left(\frac{y}{x} \right) + f_n(x, y) \right) = -F(x, y). \tag{15.12}$$

For all $x > 0$ and $0 < y \leq x$, the series

$$\sum_{n=0}^{\infty} \frac{s_n(x)}{x} f_n(x, y)$$

converges, being the inner product of $S(x, \cdot)$ and $F(\cdot, y)$ in $L_2(0, x)$. Indeed, since the set of the Legendre polynomials $\left\{ P_{2n+1} \left(\frac{\cdot}{x} \right) \right\}_{n=0}^{\infty}$ is complete and orthogonal in $L_2(0, x)$, together with (15.4) we have that

$$F(t, y) = \sum_{n=0}^{\infty} \left\langle F(\cdot, y), P_{2n+1} \left(\frac{\cdot}{x} \right) \right\rangle \frac{P_{2n+1} \left(\frac{t}{x} \right)}{\left\| P_{2n+1} \left(\frac{\cdot}{x} \right) \right\|_{L_2(0,x)}^2}$$

$$= \frac{1}{x} \sum_{n=0}^{\infty} (4n + 3) f_n(x, y) P_{2n+1} \left(\frac{t}{x} \right).$$

Consequently (due to the general Parseval identity [6, p. 16]),

$$\langle G(x, \cdot), F(\cdot, y) \rangle_{L_2(0,x)} = \sum_{n=0}^{\infty} \left\langle G(x, \cdot), \frac{P_{2n+1} \left(\frac{\cdot}{x} \right)}{\left\| P_{2n+1} \left(\frac{\cdot}{x} \right) \right\|} \right\rangle \left\langle F(\cdot, y), \frac{P_{2n+1} \left(\frac{\cdot}{x} \right)}{\left\| P_{2n+1} \left(\frac{\cdot}{x} \right) \right\|} \right\rangle$$

$$= \sum_{n=0}^{\infty} \frac{s_n(x)}{x} f_n(x, y).$$

Now, multiplying (15.12) by $P_{2m+1} \left(\frac{y}{x} \right)$ and integrating with respect to y yields

$$\frac{s_m(x)}{4m + 3} + \sum_{n=0}^{\infty} \frac{s_n(x)}{x} \int_0^x P_{2m+1} \left(\frac{y}{x} \right) f_n(x, y) dy = -f_m(x, x)$$

for all $m = 0, 1, \ldots$. A simple calculation with the aid of [141, formula 2.17.7.1] shows that

$$\frac{1}{x} \int_0^x P_{2m+1} \left(\frac{y}{x} \right) f_n(x, y) = B_{m,n}(x)$$

and hence (15.7) is obtained. The series $\sum_{n=0}^{\infty} s_n(x)B_{m,n}(x)$ converges, for all $x > 0$ again due to the general Parseval identity, as the inner product of $S(x, \cdot)$ and $f_m(x, \cdot)$ in $L_2(0, x)$. ∎

15.3 Numerical Realization and Illustration

Theorem 15.1 together with equality (15.6) lead to a simple and direct numerical algorithm for solving the inverse quantum scattering problem on a half-line. Given the spectral data,

(i) compute approximations of $f_m(x, x)$ and $B_{m,n}(x)$ by formulas (15.8) and (15.9);
(ii) for a set of points $\{x_l\}$ from $(0, b]$, $b > 0$, solve the system

$$\frac{s_m(x)}{4m + 3} + \sum_{n=0}^{M} s_n(x)B_{m,n}(x) = -f_m(x, x), \quad m = 0, \ldots, M. \tag{15.13}$$

obtaining $s_0(x)$;
(iii) compute $q(x)$, $x \in (0, b)$ by formula (15.6).

The computation of the integrals in (15.8) and (15.9) in step (i) as well as the differentiation of $s_0(x)$ in step (ii) were performed as explained in Sect. 14.3.

Example 15.1 Consider the test problem from Example 14.1,

$$q(x) = \frac{1200e^{10x}}{(6e^{10x} - 1)^2}, \quad x > 0.$$

The Jost solution is given by (see [7])

$$e(\rho, x) = e^{i\rho x}\left(\frac{10i}{\rho + 5i}\frac{1}{6e^{10x} - 1}\right), \quad x > 0.$$

Hence,

$$F(\rho) = \frac{\rho + 7i}{\rho + 5i},$$

and the discrete spectrum is empty.

With 1201 points $\{x_l\}$ uniformly distributed in the interval $[0, \pi]$, the maximum error of approximation of the potential q by the recovered potential denoted by q_M is reported in Table 15.1. Here M corresponds to the number of equations in the truncated system (15.7).

Table 15.1 Maximum absolute error of approximation of the potential q from Example 15.1 by the recovered potential

M	$\max_{x \in (0,\pi)} |q(x) - q_M(x)|$
0	0.165
1	0.016
2	0.0037
3	0.0012
4	2.19×10^{-4}
5	1.31×10^{-5}
6	6.5×10^{-6}
7	5.8×10^{-6}

Thus, again we observe that even a reduced number of equations in the truncated system (15.13) is sufficient for recovering the potential with a satisfactory accuracy.

Solution of the Inverse Scattering Problem on the Line

16.1 The Gelfand-Levitan-Marchenko Equation

In this chapter we consider the inverse scattering problem on the line introduced in Chap. 6. Given a set of scattering data J^+ or J^- as in (6.3), find the potential $q(x)$ satisfying the condition (6.2).

Denote

$$R^{\pm}(x) := \frac{1}{2\pi} \int_{-\infty}^{\infty} s^{\pm}(\rho) e^{\pm i \rho x} d\rho.$$

For definiteness let us consider the left scattering data, that is, J^-. Then the solution of the corresponding inverse scattering problem can be reduced to the following *Gelfand-Levitan-Marchenko equation*

$$F(x+y) + B(x, y) + \int_{-\infty}^{x} B(x, t) F(t+y) dt = 0, \quad y < x, \tag{16.1}$$

where $B(x, y)$ is the kernel from the Levin representation for the Jost solution $g(\rho, x)$ (see Sect. 8.3) and

$$F(x) = R^-(x) + \sum_{k=1}^{N} \alpha_k^- e^{\tau_k x}. \tag{16.2}$$

V. V. Kravchenko, *Direct and Inverse Sturm-Liouville Problems*, Frontiers in Mathematics, https://doi.org/10.1007/978-3-030-47849-0_16

For each fixed x, Eq. (16.1) possesses a unique solution $B(x, \cdot) \in L_2(-\infty, x)$, and the potential q is recovered by the formula

$$q(x) = 2\frac{d}{dx}B(x, x).$$

16.2   A Fourier–Laguerre Series Expansion of the Transmutation Kernel B

Here we briefly repeat the procedure from Sect. 10.1, applying it now to the kernel $B(x, t)$. Denote

$$b(x, t) := e^{\frac{t}{2}} B(x, x - t).$$

Then $b(x, \cdot) \in L_2\left(0, \infty; e^{-t}\right)$. Indeed,

$$\int_0^\infty |b(x, t)|^2 e^{-t} dt = \int_{-\infty}^x |B(x, t)|^2 dt < \infty.$$

Consequently, $b(x, t)$ admits the representation

$$b(x, t) = \sum_{n=0}^\infty b_n(x) L_n(t),$$

and then

$$B(x, y) = \sum_{n=0}^\infty b_n(x) L_n(x - y) e^{-\frac{x-y}{2}}. \tag{16.3}$$

To write the differential equations satisfied by the coefficients b_n, we substitute the representation (16.3) into (8.17) and obtain

$$g(\rho, x) = e^{-i\rho x}\left(1 + \int_0^\infty B(x, x - t)e^{i\rho t} dt\right)$$

$$= e^{-i\rho x}\left(1 + \int_0^\infty b(x, t)e^{-\left(\frac{1}{2} - i\rho\right)t} dt\right)$$

$$= e^{-i\rho x}\left(1 + \sum_{n=0}^\infty b_n(x) \int_0^\infty L_n(t)e^{-\left(\frac{1}{2} - i\rho\right)t} dt\right).$$

According to [76, formula 7.414 (2)],

$$\int_0^\infty L_n(t) e^{-\left(\frac{1}{2}-i\rho\right)t} dt = \frac{(-1)^n \left(\frac{1}{2}+i\rho\right)^n}{\left(\frac{1}{2}-i\rho\right)^{n+1}}.$$

Hence,

$$g(\rho, x) = e^{-i\rho x} \left(1 + \sum_{n=0}^\infty \frac{(-1)^n \left(\frac{1}{2}+i\rho\right)^n}{\left(\frac{1}{2}-i\rho\right)^{n+1}} b_n(x)\right).$$

Substitution of this expression into Eq. (6.1) leads to the equality

$$q(x) \left(1 + \sum_{n=0}^\infty \frac{(-1)^n \left(\frac{1}{2}+i\rho\right)^n}{\left(\frac{1}{2}-i\rho\right)^{n+1}} b_n(x)\right) \tag{16.4}$$

$$= \sum_{n=0}^\infty \frac{(-1)^n \left(\frac{1}{2}+i\rho\right)^n}{\left(\frac{1}{2}-i\rho\right)^{n+1}} b_n''(x) - 2i\rho \sum_{n=0}^\infty \frac{(-1)^n \left(\frac{1}{2}+i\rho\right)^n}{\left(\frac{1}{2}-i\rho\right)^{n+1}} b_n'(x).$$

Again it is convenient to introduce the parameter (10.8), $z := \left(\frac{1}{2}+i\rho\right) / \left(\frac{1}{2}-i\rho\right)$. Then $i\rho = (z-1)/(2(z+1))$ and $\frac{1}{2}-i\rho = 1/(z+1)$. The Jost solution $g(\rho, x)$ takes the form (compare to the series representation (10.9) of $e(\rho, x)$)

$$g(\rho, x) = e^{-i\rho x} \left(1 + (z+1) \sum_{n=0}^\infty (-1)^n z^n b_n(x)\right).$$

Therefore, (16.4) can be written in the form

$$q(x) + q(x) (z+1) \sum_{n=0}^\infty (-1)^n z^n b_n(x)$$

$$= (z+1) \sum_{n=0}^\infty (-1)^n z^n b_n''(x) - (z-1) \sum_{n=0}^\infty (-1)^n z^n b_n'(x).$$

Equating the terms corresponding to identical powers of z, we obtain the equalities

$$b_0'' + b_0' - q b_0 = q \tag{16.5}$$

and

$$Lb_n + b'_n = Lb_{n-1} - b'_{n-1}, \quad n = 1, 2, \ldots,$$

where $L := \frac{d^2}{dx^2} - q(x)$. Note that (16.5) implies that

$$q = \frac{b''_0 + b'_0}{b_0 + 1}. \tag{16.6}$$

Thus, the potential q can be recovered from the first coefficient of the Fourier-Laguerre series of the kernel B, and there is no need to compute the whole kernel.

Now let us obtain a system of linear algebraic equations for the coefficients $\{b_n\}_{n=0}^{\infty}$. Substitution of (16.3) into (16.1) gives

$$F(x + y) + \sum_{n=0}^{\infty} b_n(x) L_n(x - y) e^{-\frac{x-y}{2}}$$

$$+ \sum_{n=0}^{\infty} b_n(x) \int_{-\infty}^{x} L_n(x - t) e^{-\frac{x-t}{2}} F(t + y) dt = 0, \quad y < x,$$

which then can be written as

$$F(x + y) + \sum_{n=0}^{\infty} B_n(x) L_n(x - y) e^{-\frac{x-y}{2}}$$

$$+ \sum_{n=0}^{\infty} b_n(x) \int_{0}^{\infty} L_n(t) e^{-\frac{t}{2}} F(x + y - t) dt = 0.$$

Denote

$$f_n(x) := \int_{0}^{\infty} L_n(t) e^{-\frac{t}{2}} F(x - t) dt.$$

Thus,

$$F(x + y) + \sum_{n=0}^{\infty} b_n(x) L_n(x - y) e^{-\frac{x-y}{2}} + \sum_{n=0}^{\infty} b_n(x) f_n(x + y) = 0.$$

Let $s := x - y$. Then

$$F(2x - s) + \sum_{n=0}^{\infty} b_n(x) L_n(s) e^{-\frac{s}{2}} + \sum_{n=0}^{\infty} b_n(x) f_n(2x - s) = 0.$$

Multiplication of this equality by $L_m(s)e^{-\frac{s}{2}}$ and integration leads to the system of equations

$$b_m(x) + \sum_{n=0}^{\infty} b_n(x)B_{mn}(x) = -f_m(2x), \quad m = 0, 1, 2, \ldots, \tag{16.7}$$

where

$$B_{mn}(x) := \int_0^{\infty} f_n(2x - s)L_m(s)e^{-\frac{s}{2}}ds.$$

16.3 Expressions for the Coefficients of the System

Let us obtain expressions for the functions f_n and B_{mn} that are convenient for numerical computation. Since they are determined by the function F, which in its turn consists of two parts (see (16.2)) which can be called discrete and continuous, it is useful to consider them separately. Thus,

$$f_n = f_n^d + f_n^c \tag{16.8}$$

where

$$f_n^d(x) := \sum_{k=1}^{N} \alpha_k \int_0^{\infty} L_n(t)e^{-\frac{t}{2}}e^{\tau_k(x-t)}dt,$$

$$f_n^c(x) := \frac{1}{2\pi} \int_0^{\infty} L_n(t)e^{-\frac{t}{2}} \int_{-\infty}^{\infty} s^-(\rho) e^{-i\rho(x-t)}d\rho dt,$$

and

$$B_{mn} = B_{mn}^d + B_{mn}^c \tag{16.9}$$

where

$$B_{mn}^{d,c}(x) := \int_0^{\infty} f_n^{d,c}(2x - s)L_m(s)e^{-\frac{s}{2}}ds.$$

We begin with the discrete components. We have

$$f_n^{\mathrm{d}}(x) = \sum_{k=1}^{N} \alpha_k e^{\tau_k x} \int_0^\infty L_n(t) e^{-\left(\frac{1}{2}+\tau_k\right)t}\,dt$$

$$= \sum_{k=1}^{N} \alpha_k e^{\tau_k x} \frac{\left(-\frac{1}{2}+\tau_k\right)^n}{\left(\frac{1}{2}+\tau_k\right)^{n+1}} \tag{16.10}$$

and hence

$$B_{mn}^{\mathrm{d}}(x) = \sum_{k=1}^{N} \alpha_k e^{2\tau_k x} \frac{\left(-\frac{1}{2}+\tau_k\right)^n}{\left(\frac{1}{2}+\tau_k\right)^{n+1}} \int_0^\infty L_m(s) e^{-\left(\frac{1}{2}+\tau_k\right)s}\,ds$$

$$= \sum_{k=1}^{N} \alpha_k e^{2\tau_k x} \frac{\left(-\frac{1}{2}+\tau_k\right)^{n+m}}{\left(\frac{1}{2}+\tau_k\right)^{n+m+2}}. \tag{16.11}$$

For the continuous components we have

$$f_n^{\mathrm{c}}(x) = \frac{1}{2\pi} \int_{-\infty}^\infty s^-(\rho) e^{-i\rho x} \int_0^\infty L_n(t) e^{-\frac{t}{2}} e^{i\rho t}\,dt\,d\rho$$

$$= \frac{(-1)^n}{2\pi} \int_{-\infty}^\infty s^-(\rho) e^{-i\rho x} \frac{\left(\frac{1}{2}+i\rho\right)^n}{\left(\frac{1}{2}-i\rho\right)^{n+1}}\,d\rho \tag{16.12}$$

and

$$B_{mn}^{\mathrm{c}}(x) = \int_0^\infty f_n^{\mathrm{c}}(2x-s) L_m(s) e^{-\frac{s}{2}}\,ds$$

$$= \frac{(-1)^n}{2\pi} \int_{-\infty}^\infty s^-(\rho) e^{-2i\rho x} \frac{\left(\frac{1}{2}+i\rho\right)^n}{\left(\frac{1}{2}-i\rho\right)^{n+1}} \int_0^\infty L_m(s) e^{-\frac{s}{2}} e^{i\rho s}\,ds\,d\rho$$

$$= \frac{(-1)^{n+m}}{2\pi} \int_{-\infty}^\infty s^-(\rho) \frac{\left(\frac{1}{2}+i\rho\right)^{n+m}}{\left(\frac{1}{2}-i\rho\right)^{n+m+2}} e^{-2i\rho x}\,d\rho. \tag{16.13}$$

16.4 Numerical Realization

The following numerical method for solving the inverse scattering problem can be proposed.

1. Given the set of the left scattering data J^-. Compute $f_m(2x)$ and $B_{mn}(x), m, n = \overline{0, N_s}$ using formulas (16.8)–(16.13).
2. Solve the system (16.7).
3. Compute q by formula (16.6).

For the first numerical illustration we consider a simple example of a reflectionless potential. The reflectionless potentials are characterized by the property $s^{\pm} \equiv 0$ and hence $F(x) = \sum_{k=1}^{N} \alpha_k e^{\tau_k x}$.

Thus, in the case of a reflectionless potential the approximate system obtained from (16.7) has the form

$$b_m(x) + \sum_{n=0}^{N_s} b_n(x) \sum_{k=1}^{N} \alpha_k e^{2\tau_k x} \frac{\left(-\frac{1}{2} + \tau_k\right)^{n+m}}{\left(\frac{1}{2} + \tau_k\right)^{n+m+2}} = -\sum_{k=1}^{N} \alpha_k e^{2\tau_k x} \frac{\left(-\frac{1}{2} + \tau_k\right)^{m}}{\left(\frac{1}{2} + \tau_k\right)^{m+1}},$$

$$m = 0, \dots, N_s. \qquad (16.14)$$

Example 16.1 Let α and τ be two arbitrary positive numbers. Then using a simple procedure described in [172] it is easy to find the reflectionless potential q with the scattering data $J^- = \{s^-(\rho) \equiv 0, \rho \in \mathbf{R}; \lambda_1 = -\tau^2, \alpha_1^- = \alpha, N = 1\}$. Namely,

$$q(x) = -\frac{2}{\alpha} \frac{(2\tau)^3 e^{2\tau x}}{\left(\frac{2\tau}{\alpha} + e^{2\tau x}\right)^2}. \qquad (16.15)$$

Figure 16.1 shows the recovered potential (marked with circles) in comparison with the exact potential in the case $\alpha = 1$, $\tau = 2$ and $N_s = 10$. The absolute error is of order 10^{-2}. To use formula (16.6) the function $b_0(x)$ obtained by solving the system (16.14) was approximated by a spline of sixth order and differentiated twice.

Taking $N_s = 20$ reduces the absolute error to the order 10^{-5}.

The second example is from [8].

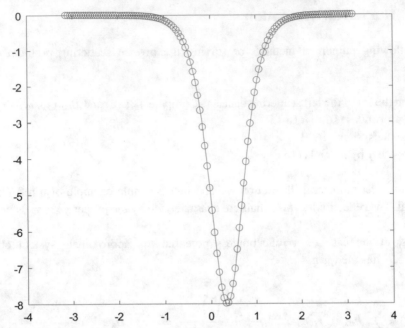

Fig. 16.1 The recovered reflectionless potential represented by circle markers compared to the exact potential (16.15) represented by the line plot. Here $N_s = 10$

Example 16.2 Let

$$s^+(\rho) = \frac{(\rho + i)(\rho + 2i)(101\rho^2 - 3i\rho - 400)}{(\rho - i)(\rho - 2i)(50\rho^4 + 280i\rho^3 - 609\rho^2 - 653i\rho + 400)}.$$

The unique potential without bound states for this reflection coefficient has the form

$$q(x) = \begin{cases} q_1(x), & \text{if } x < 0, \\ q_2(x), & \text{if } x > 0, \end{cases}$$

where

$$q_1(x) = \frac{16\left(\sqrt{2} + 1\right)^2 e^{-2\sqrt{2}x}}{\left(\left(\sqrt{2} + 1\right)^2 e^{-2\sqrt{2}x} - 1\right)^2}$$

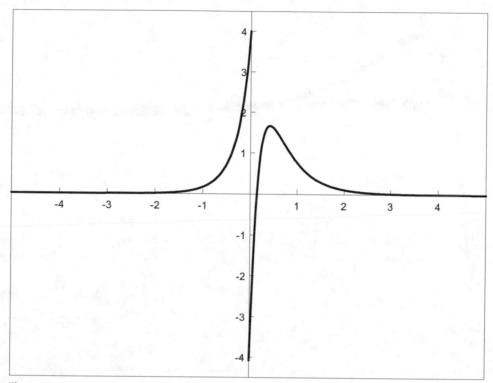

Fig. 16.2 The potential $q(x)$ from Example 16.2

and

$$q_2(x) = \frac{96e^{2x}\left(81e^{8x} - 144e^{6x} + 54e^{4x} - 9e^{2x} + 1\right)}{\left(36e^{6x} - 27e^{4x} + 12e^{2x} - 1\right)^2}.$$

The graph of this discontinuous potential $q(x)$ is presented on Fig. 16.2.

In the numerical examples presented below we computed separately the recovered potential for $x < 0$ and $x > 0$, considering the truncated system (16.7) for the intervals $-\pi < x < 0$ and $0 < x < \pi$, respectively, with 201 points $\{x_l\}$ uniformly distributed in both cases. The differentiation of $b_0(x)$ at step (iii) was performed by fitting the computed $b_0(x)$ with a partial sum of a Fourier series containing eight elements (the maximum allowed by the Matlab routine 'fit') and applying the Matlab command 'differentiate'. This produced better results than the use of a spline, as was done in the previous numerical examples.

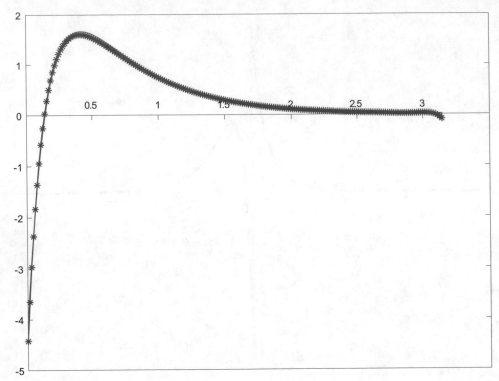

Fig. 16.3 With one equation only, the recovered potential approximates the exact $q_2(x)$ with the maximal absolute error 0.344 achieved at the origin

Already with one equation in the truncated system we obtain a reasonably good approximation of the potential $q_2(x)$, presented on Fig. 16.3.

With only five equations, the recovered potential $q_2(x)$ practically coincides with the exact one (Fig. 16.4).

For an accurate recovery of $q_1(x)$ more equations were necessary. Figure 16.5 shows the result obtained with 15 equations in the truncated system. The maximum absolute error in this case was 0.378 achieved in the origin.

With 25 equations in the truncated system the recovered potential is presented in Fig. 16.6. The maximum absolute error in this case was 0.097, achieved at the origin.

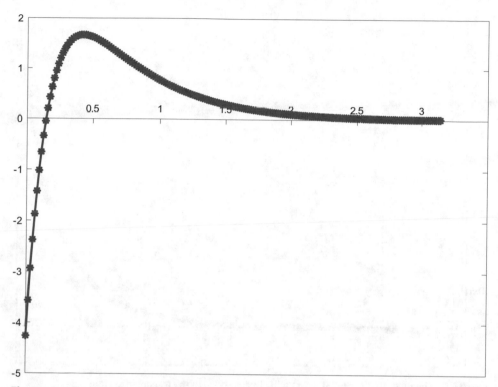

Fig. 16.4 With only five equations, the recovered potential $q_2(x)$ practically coincides with the exact one

We conclude this part of the book by commenting that the approach presented here for solving inverse spectral problems is direct, simple in its numerical realization, and allows one to recover the potential from the spectral or scattering data quickly and accurately. It was proposed first in [103] and [104] and developed further in [58] and [85]. Inverse spectral problems are, in general, computationally challenging, especially in the case of the problems on infinite intervals. Hence, the mere existence of a universal, simple and direct approach for their solution is surprising. However, this advancement is based in good measure on a well-developed and beautiful theory and thus stands on the shoulders of giants, such as Ch. Sturm, J. Liouville, H. Weyl, D. Hilbert, V. A. Steklov, J. Delsarte, G. Borg, E. Ch. Titchmarsh, A. Ya. Povzner, I. M. Gelfand, B. M. Levitan, M. G. Krein, V. A. Marchenko, B. Ya. Levin and many other brilliant mathematicians and physicists who contributed to its creation.

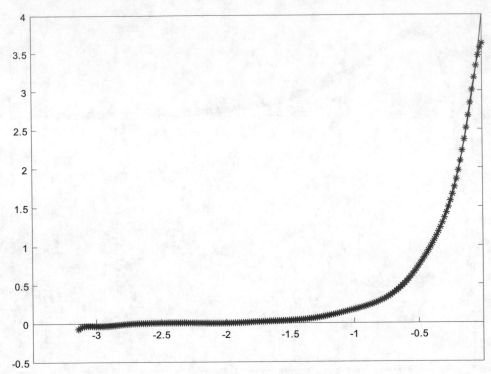

Fig. 16.5 The recovered potential $q_1(x)$ with 15 equations in the truncated system

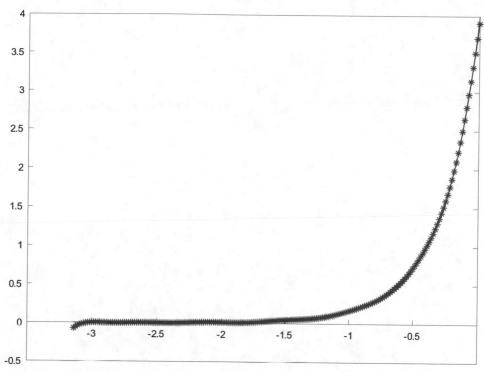

Fig. 16.6 The recovered potential $q_1(x)$ with 25 equations in the truncated system

Bibliography

1. M.J. Ablowitz, *Nonlinear Dispersive Waves: Asymptotic Analysis and Solitons* (Cambridge University Press, Cambridge, 2011)
2. M.J. Ablowitz, P. Clarkson, *Solitons, Nonlinear Evolution Equations and Inverse Scattering* (Cambridge University Press, Cambridge, 1991)
3. M.J. Ablowitz, H. Segur, *Solitons and the Inverse Scattering Transform* (Society for Industrial and Applied Mathematics (SIAM), Philadelphia, 2000)
4. M.J. Ablowitz, D.J. Kaup, A.C. Newell, H. Segur, The inverse scattering transform—Fourier analysis for nonlinear problems. Stud. Appl. Math. **53**, 249–315 (1974)
5. M. Abramovitz, I.A. Stegun, *Handbook of Mathematical Functions* (Dover, New York, 1972)
6. N.I. Akhiezer, I.M. Glazman,*Theory of Linear Operators in Hilbert Space* (Dover, New York, 1993)
7. T. Aktosun, Construction of the half-line potential from the Jost function. Inverse Prob. **20**, 859–876 (2004)
8. T. Aktosun, P. Sacks, Potential splitting and numerical solution of the inverse scattering problem on the line. Math. Methods Appl. Sci. **25**(4), 347–355 (2002)
9. T. Aktosun, P. Sacks, M. Unlu, Inverse problems for selfadjoint Schrödinger operator on the half line with compactly supported potentials. J. Math. Phys. **56**, 022106 (2015)
10. M. Al-Gwaiz, *Sturm–Liouville Theory and Its Applications* (Springer, London, 2008)
11. A.R. Aliev, S.G. Gasymova, D.G. Gasymova, N.D. Ahmadzadeh, Approximate construction of the Jost function by the collocation method for Sturm–Liouville boundary value problem. Azerbaijan J. Math. **3**(2), 45–61 (2013)
12. H. Altundag, C. Böckmann, H. Taseli, Inverse Sturm–Liouville problems with pseudospectral methods. Int. J. Comput. Math. **92**, 1373–1384 (2015)
13. V.A. Ambartsumyan, Über eine frage der eigenwerttheorie. Z. Phys. **53**, 690–695 (1929)
14. A.L. Andrew, Numerical solution of inverse Sturm–Liouville problems. Anziam J. **45**(C), 326–337 (2004)
15. A.L. Andrew, Numerov's method for inverse Sturm–Liouville problems. Inverse Prob. **21**, 223–238 (2005)
16. A.L. Andrew, Finite difference methods for half inverse Sturm–Liouville problems. Appl. Math. Comput. **218**, 445–457 (2011)
17. A. Baricz, D. Jankov, T.K. Pogány, Neumann series of Bessel functions. Integral Transforms Spec. Funct. **23**(7), 529–538 (2012)
18. A. Baricz, D. Jankov, T.K. Pogány, *Series of Bessel and Kummer-Type Functions*. Lecture Notes in Mathematics, vol. 2207 (Springer, Cham, 2017)

© The Editor(s) (if applicable) and The Author(s), under exclusive licence to Springer
Nature Switzerland AG 2020
V. V. Kravchenko, *Direct and Inverse Sturm-Liouville Problems*,
Frontiers in Mathematics, https://doi.org/10.1007/978-3-030-47849-0

19. V. Barrera-Figueroa, Analysis of the spectral singularities of Schrödinger operator with complex potential by means of the SPPS method. J. Phys. Conf. Series **698**, 012029 (2016)

20. V. Barrera-Figueroa, A power series representation for the characteristic equation of Gamow–Siegert eigenstates. J. Phys. Conf. Series **839**(1), 012004 (2017)

21. V. Barrera-Figueroa, H. Blancarte, V.V. Kravchenko, The phase retrieval problem: a spectral parameter power series approach. J. Eng. Math. **85**(1), 179–209 (2014)

22. V. Barrera-Figueroa, V.V. Kravchenko, V.S. Rabinovich, Spectral parameter power series analysis of isotropic planarly layered waveguides. Appl. Anal. **93**(4), 729–755 (2014)

23. V. Barrera-Figueroa, V.S. Rabinovich, Asymptotics of the far field generated by a modulated point source in a planarly layered electromagnetic waveguide. Math. Methods Appl. Sci. **38**(10), 1970–1989 (2015)

24. V. Barrera-Figueroa, V.S. Rabinovich, Electromagnetic field generated by a modulated moving point source in a planarly layered waveguide. Russ. J. Math. Phys. **23**(2), 139–163 (2016)

25. V. Barrera-Figueroa, V.S. Rabinovich, Effective numerical method of spectral analysis of quantum graphs. *J. Phys. A Math. Theor.* **50**(21), 215207 (2017)

26. V. Barrera-Figueroa, V.S. Rabinovich, Numerical calculation of the discrete spectra of one-dimensional Schrödinger operators with point interactions. Math. Methods Appl. Sci. **42**(15), 5072–5093 (2019)

27. V. Barrera-Figueroa, V.S. Rabinovich, M. Maldonado Rosas, Numerical estimates of the essential spectra of quantum graphs with delta-interactions at vertices. Appl. Anal. **98**(1–2), 458–482 (2019)

28. H. Begehr, R. Gilbert, *Transformations, Transmutations and Kernel Functions*, vol. 1–2 (Longman Scientific & Technical, Harlow, 1992)

29. R. Bellman, *Perturbation Techniques in Mathematics, Engineering and Physics* (Dover Publications, New York, 2003)

30. J. Ben Amara, A.A. Shkalikov, A Sturm–Liouville problem with physical and spectral parameters in boundary conditions. Math. Notes **66**(2), 127–134 (1999)

31. J. Behrndt, H. de Snoo, S. Hassi, *Boundary Value Problems, Weyl Functions, and Differential Operators* (Birkhäuser, Basel, 2020)

32. H. Blancarte, H.M. Campos, K.V. Khmelnytskaya, Spectral parameter power series method for discontinuous coefficients. Math. Methods Appl. Sci. **38**, 2000–2011 (2015)

33. G. Borg, Eine Umkehrung der Sturm–Liouville Eigenwertaufgabe. Acta Math. **76**, 1–96 (1946)

34. A. Boumenir, The approximation of the transmutation kernel. J. Math. Phys. **47**, 013505 (2006)

35. B.M. Brown, V.S. Samko, I.W. Knowles, M. Marletta, Inverse spectral problem for the Sturm–Liouville equation. Inverse Prob. **19**, 235–252 (2003)

36. Ph.R. Brown, R.M. Porter, Conformal mapping of circular quadrilaterals and Weierstrass elliptic functions. Comput. Methods Funct. Theory **11**(2), 463–486 (2012)

37. R. Camporesi, A.J. Di Scala, A generalization of a theorem of Mammana. Colloq. Math. **122**(2), 215–223 (2011)

38. H.M. Campos, V.V. Kravchenko, A finite-sum representation for solutions for the Jacobi operator. J. Differ. Equ. Appl. **17**(4), 567–575 (2011)

39. H. Campos, R. Castillo-Pérez, V.V. Kravchenko, Construction and application of Bergman-type reproducing kernels for boundary and eigenvalue problems in the plane. Complex Var. Elliptic Equ. **57**(7–8), 787–824 (2012)

40. H. Campos, V.V. Kravchenko, L.M. Méndez, Complete families of solutions for the Dirac equation: an application of bicomplex pseudoanalytic function theory and transmutation operators. Adv. Appl. Clifford Algebr. **22**(3), 577–594 (2012)

41. H. Campos, V.V. Kravchenko, S.M. Torba, Transmutations, L-bases and complete families of solutions of the stationary Schrödinger equation in the plane. J. Math. Anal. Appl. **389**, 1222–1238 (2012)

42. R.W. Carroll, *Transmutation and Operator Differential Equations*. Mathematics Studies, vol. 37 (North Holland, Amsterdam, 1979)
43. R.W. Carroll, *Transmutation, Scattering theory and Special Functions*. Mathematics Studies, vol. 69 (North Holland, Amsterdam, 1982)
44. R.W. Carroll, *Transmutation Theory and Applications*. Mathematics Studies, vol. 117 (North-Holland, Amsterdam, 1985)
45. R. Castillo-Pérez, K.V. Khmelnytskaya, V.V. Kravchenko, H. Oviedo-Galdeano, Efficient calculation of the reflectance and transmittance of finite inhomogeneous layers. J. Optics A Pure Appl. Optics **11**(6), 065707 (2009)
46. R. Castillo-Pérez, V.V. Kravchenko, H. Oviedo, V.S. Rabinovich, Dispersion equation and eigenvalues for quantum wells using spectral parameter power series. J. Math. Phys. **52**(4), 043522 (2011)
47. R. Castillo-Pérez, V.V. Kravchenko, S.M. Torba, Spectral parameter power series for perturbed Bessel equations. Appl. Math. Comput. **220**(1), 676–694 (2013)
48. R. Castillo-Pérez, V.V. Kravchenko, S.M. Torba, Analysis of graded-index optical fibers by the spectral parameter power series method. J. Optics **17**, 025607 (2015)
49. R. Castillo-Pérez, V.V. Kravchenko, S.M. Torba, A method for computation of scattering amplitudes and Green functions of whole axis problems. Math. Methods Appl. Sci. **42**(15), 5106–5117 (2019)
50. Kh. Chadan, D. Colton, L. Päivärinta, W. Rundell, *An Introduction to Inverse Scattering and Inverse Spectral Problems* (SIAM, Philadelphia, 1997)
51. Kh. Chadan, P.C. Sabatier, *Inverse Problems in Quantum Scattering Theory* (Springer, Berlin, 1989)
52. B. Chanane, Sturm–Liouville problems with parameter dependent potential and boundary conditions. J. Comput. Appl. Math. **212**(2), 282–290 (2008)
53. H. Chébli, A. Fitouhi, M.M. Hamza, Expansion in series of Bessel functions and transmutations for perturbed Bessel operators. J. Math. Anal. Appl. **181**(3), 789–802 (1994)
54. M.S. Child, A.V. Chambers, Persistent accidental degeneracies for the Coffey-Evans potential. J. Phys. Chem. **92**, 3122–3124 (1988)
55. W.J. Code, P.J. Browne, Sturm–Liouville problems with boundary conditions depending quadratically on the eigenparameter. J. Math. Anal. Appl. **309**(2), 729–742 (2005)
56. D.L. Colton, *Solution of Boundary Value Problems by the Method of Integral Operators* (Pitman, London, 1976)
57. H. Coşkun, N. Bayram, Asymptotics of eigenvalues for regular Sturm–Liouville problems with eigenvalue parameter in the boundary condition. J. Math. Anal. Appl. **306**(2), 548–566 (2005)
58. B.B. Delgado, K.V. Khmelnytskaya, V.V. Kravchenko, The transmutation operator method for efficient solution of the inverse Sturm–Liouville problem on a half-line. Math. Methods Appl. Sci. **42**(18), 7359–7366 (2019)
59. B.B. Delgado, K.V. Khmelnytskaya, V.V. Kravchenko, A representation for Jost solutions and an efficient method for solving the spectral problem on the half line. Math. Methods Appl. Sci. (2019). https://doi.org/10.1002/mma.5881
60. J. Delsarte, Sur une extension de la formule de Taylor. J Math. Pures et Appl. **17**, 213–230 (1938)
61. J. Delsarte, Sur certaines transformations fonctionnelles relatives aux équations linéaires aux dérivé es partielles du second ordre. C. R. Acad. Sci. **206**,178–182 (1938)
62. J. Delsarte, J.L. Lions, Transmutations d'opérateurs différentiels dans le domaine complexe. Comment. Math. Helv. **32**, 113–128 (1956)
63. L. Erbe, R. Mert, A. Peterson, Spectral parameter power series for Sturm–Liouville equations on time scales. Appl. Math. Comput. **218**(14), 7671–7678 (2012)

64. W.N. Everitt, A catalogue of Sturm–Liouville differential equations, in *Sturm–Liouville Theory, Past and Present* (Birkhäuser, Basel, 2005), pp. 271–331
65. M.K. Fage, N.I. Nagnibida, *The Problem of Equivalence of Ordinary Linear Differential Operators* (Nauka, Novosibirsk, 1987) (in Russian)
66. M.V. Fedoryuk, *Ordinary Differential Equations* (URSS, Moscow, 2017) (in Russian)
67. A. Fitouhi, M.M. Hamza, A uniform expansion for the eigenfunction of a singular second-order differential operator. *SIAM J. Math. Anal.* **21**(6), 1619–1632 (1990)
68. G. Freiling, T. Mazur, V. Yurko, A numerical algorithm for solving inverse problems for singular Sturm–Liouville operators. Adv. Dyn. Syst. Appl. **2**(1), 95–105 (2007)
69. G. Freiling, V. Yurko, *Inverse Sturm–Liouville Problems and Their Applications* (Nova Science Publishers, Huntington, 2001)
70. C. Fulton, Two-point boundary value problems with eigenvalue parameter contained in the boundary conditions. Proc. R. Soc. Edinburgh Sect. A **77**(3–4), 293–308 (1977)
71. C. Fulton, D. Pearson, S. Pruess, Computing the spectral function for singular Sturm–Liouville problem. J. Comput. Appl. Math. **176**, 131–162 (2005)
72. C. Fulton, D. Pearson, S. Pruess, Efficient calculation of spectral density functions for specific classes of singular Sturm–Liouville problems. J. Comput. Appl. Math. **212**, 150–178 (2008)
73. C.S. Gardner, J.M. Greene, M.D. Kruskal, R. M. Miura, Method for solving the Korteweg-de Vries equation. Phys. Rev. Lett., **19**(19), 1095–1097 (1967)
74. I.M. Gelfand, B.M. Levitan, On the determination of a differential equation from its spectral function. Izvestiya AN SSSR, Ser. matem. **15**(4), 309–360 (1951)
75. K. Gou, Z. Chen, Inverse Sturm–Liouville problems and their biomedical engineering applications. J. SciMed. Central Math. Stat. **2**(1), 1008 (2015)
76. I. Gradshteyn, I. Ryzhik, *Table of Integrals, Series, and Products* (Academic Press, New York, 1980)
77. N. Gutiérrez Jiménez, S.M. Torba, Spectral parameter power series representation for solutions of linear system of two first order differential equations. Appl. Math. Comput. **370**, 124911 (2020)
78. Zh. Han, Y. Hu, Ch. Lee, Optimal pricing barriers in a regulated market using reflected diffusion processes. Quant. Financ. **16**(4), 639–647 (2016)
79. J. Hernández-Juárez, D.A. Serrano, A. López-Villa, A. Medina, A new methodology in the study of acoustic fields in the almost stratified ocean. J. Phys. Conf. Series **1221**(1), 012071 (2019)
80. M. Ignatiev, V. Yurko, Numerical methods for solving inverse Sturm–Liouville problems. Results Math. **52**(1–2), 63–74 (2008)
81. D. Jackson, *The Theory of Approximation. Reprint of the 1930 Original* (American Mathematical Society, Providence, 1994)
82. S.I. Kabanikhin, *Inverse and Ill-Posed Problems: Theory and Applications* (De Gruyter, Berlin, 2012)
83. A. Kammanee, C. Böckmann, Boundary value method for inverse Sturm–Liouville problems. Appl. Math. Comput. **214**, 342–352 (2009)
84. L.V. Kantorovich, G.P. Akilov, *Functional Analysis*, 2nd edn. (Pergamon Press, New York, 1982)
85. A.N. Karapetyants, K.V. Khmelnytskaya, V.V. Kravchenko, A practical method for solving the inverse quantum scattering problem on a half line. J. Phys. Conf. Series (to appear)
86. V.V. Katrakhov, S.M. Sitnik, The transmutation method and boundary value problems for singular elliptic equations. Contemp. Math. Fundam. Dir. **64**(2), 211–426 (2018) (in Russian)
87. K.V. Khmelnytskaya, V.V. Kravchenko, J.A. Baldenebro-Obeso, Spectral parameter power series for fourth-order Sturm–Liouville problems. Appl. Math. Comput. **219**(8), 3610–3624 (2012)

88. K.V. Khmelnytskaya, V.V. Kravchenko, H.C. Rosu, Eigenvalue problems, spectral parameter power series, and modern applications. Math. Methods Appl. Sci. **38**, 1945–1969 (2015)

89. K.V. Khmelnytskaya, V.V. Kravchenko, S.M. Torba, Modulated electromagnetic fields in inhomogeneous media, hyperbolic pseudoanalytic functions and transmutations. J. Math. Phys. **57**, 051503 (2016)

90. K.V. Khmelnytskaya, V.V. Kravchenko, S.M. Torba, A representation of the transmutation kernels for the Schrödinger operator in terms of eigenfunctions and applications. Appl. Math. Comput. **353**, 274–281 (2019)

91. K.V. Khmelnytskaya, V.V. Kravchenko, S.M. Torba, Time-dependent one-dimensional electromagnetic wave propagation in inhomogeneous media: exact solution in terms of transmutations and Neumann series of Bessel functions. Lobachevskii J. Math. **41**(5), 785–796 (2020)

92. K.V. Khmelnytskaya, V.V. Kravchenko, S.M. Torba, S. Tremblay, Wave polynomials, transmutations and Cauchy's problem for the Klein-Gordon equation. J. Math. Anal. Appl. **399**, 191–212 (2013)

93. K.V. Khmelnytskaya, H.C. Rosu, Spectral parameter power series representation for Hill's discriminant. Ann. Phys. **325**(11), 2512–2521 (2010)

94. K.V. Khmelnytskaya, H.C. Rosu, Bloch solutions of periodic Dirac equations in SPPS form. Oper. Theory Adv. Appl. **220**, 153–162 (2012)

95. K.V. Khmelnytskaya, I. Scrroukh, The heat transfer problem for inhomogeneous materials in photoacoustic applications and spectral parameter power series. Math. Methods Appl. Sci. **36**(14), 1878–1891 (2013)

96. K.V. Khmelnytskaya, T.V. Torchynska, Reconstruction of potentials in quantum dots and other small symmetric structures. Math. Methods Appl. Sci. **33**, 469–472 (2010)

97. A. Kostenko, G. Teschl, On the singular Weyl–Titchmarsh function of perturbed spherical Schrödinger operators. J. Differ. Equ. **250**, 3701–3739 (2011)

98. I.V. Kravchenko, V.V. Kravchenko, S.M. Torba, Solution of parabolic free boundary problems using transmuted heat polynomials. Math. Methods Appl. Sci. **42**, 5094–5105 (2019)

99. I.V. Kravchenko, V.V. Kravchenko, S.M. Torba, J.C. Dias, Generalized exponential basis for efficient solving of homogeneous diffusion free boundary problems: Russian option pricing. https://arxiv.org/abs/1808.08290

100. V.V. Kravchenko, A representation for solutions of the Sturm–Liouville equation. Complex Var. Elliptic Equ. **53**, 775–789 (2008)

101. V.V. Kravchenko, *Applied Pseudoanalytic Function Theory*. Frontiers in Mathematics (Birkhäuser, Basel, 2009)

102. V.V. Kravchenko, Construction of a transmutation for the one-dimensional Schrödinger operator and a representation for solutions. Appl. Math. Comput. **328**, 75–81 (2018)

103. V.V. Kravchenko, On a method for solving the inverse Sturm–Liouville problem. J. Inverse Ill-Posed Prob. **27**, 401–407 (2019)

104. V.V. Kravchenko, On a method for solving the inverse scattering problem on the line. Math. Methods Appl. Sci. **42**, 1321–1327 (2019)

105. V.V. Kravchenko, S. Morelos, S.M. Torba, Liouville transformation, analytic approximation of transmutation operators and solution of spectral problems. Appl. Math. Comput. **273**, 321–336 (2016)

106. V.V. Kravchenko, L.J. Navarro, S.M. Torba, Representation of solutions to the one-dimensional Schrödinger equation in terms of Neumann series of Bessel functions. Appl. Math. Comput. **314**(1), 173–192 (2017)

107. V.V. Kravchenko, J.A. Otero, S.M. Torba, Analytic approximation of solutions of parabolic partial differential equations with variable coefficients. Adv. Math. Phys. **2017**, 2947275 (2017)

108. V.V. Kravchenko, R.M. Porter, Spectral parameter power series for Sturm–Liouville problems. Math. Method Appl. Sci. **33**, 459–468 (2010)

109. V.V. Kravchenko, R.M. Porter, Conformal mapping of right circular quadrilaterals. Complex Variables Elliptic Equ. **56**(5), 399–415 (2011)

110. V.V. Kravchenko, R.M. Porter, S.M. Torba, Spectral parameter power series for arbitrary order linear differential equations. Math. Methods Appl. Sci. **42**(15), 4902–4908 (2019)

111. V.V. Kravchenko, E.L. Shishkina, S.M. Torba, On a series representation for integral kernels of transmutation operators for perturbed Bessel equations. Math. Notes **104**(3–4), 530–544 (2018)

112. V.V. Kravchenko, S.M. Sitnik (eds.), *Transmutation Operators and Applications*. Trends in Mathematics (Birkhäuser, Basel, 2020)

113. V.V. Kravchenko, S.M. Torba, Transmutations and spectral parameter power series in eigenvalue problems, in *Operator Theory, Pseudo-Differential Equations, and Mathematical Physics* (Springer, Basel, 2013), pp. 209–238

114. V.V. Kravchenko, S. Torba, Modified spectral parameter power series representations for solutions of Sturm–Liouville equations and their applications. Appl. Math. Comput. **238**, 82–105 (2014)

115. V.V. Kravchenko, S.M. Torba, Construction of transmutation operators and hyperbolic pseudo-analytic functions. Complex Anal. Oper. Theory **9**, 389–429 (2015)

116. V.V. Kravchenko, S.M. Torba, Analytic approximation of transmutation operators and applications to highly accurate solution of spectral problems. J. Comput. Appl. Math. **275**, 1–26 (2015)

117. V.V. Kravchenko, S.M. Torba, Analytic approximation of transmutation operators and related systems of functions. Bol. Soc. Mat. Mex. **22**, 379–429 (2016)

118. V.V. Kravchenko, S.M. Torba, A Neumann series of Bessel functions representation for solutions of Sturm–Liouville equations. Calcolo 55, 11 (2018)

119. V.V. Kravchenko, S.M. Torba, R. Castillo-Pérez, A Neumann series of Bessel functions representation for solutions of perturbed Bessel equations. Appl. Anal. **97**(5), 677–704 (2018)

120. V.V. Kravchenko, S.M. Torba, K.V. Khmelnytskaya, Transmutation operators: construction and applications, in *Proceedings of the 17th International Conference on Computational and Mathematical Methods in Science and Engineering CMMSE-2017*, Cadiz (2017), pp. 1198–1206

121. V.V. Kravchenko, S.M. Torba, U. Velasco-García, Spectral parameter power series for Sturm–Liouville equations with a potential polynomially dependent on the spectral parameter and Zakharov-Shabat systems. J. Math. Phys. **56**, 073508 (2015)

122. V.V. Kravchenko, U. Velasco-García, Dispersion equation and eigenvalues for the Zakharov-Shabat system using spectral parameter power series. J. Math. Phys. **52**(6), 063517 (2011)

123. M.M. Lavrent'ev, K.G. Reznitskaya, V.G. Yakhno, *One-Dimensional Inverse Problems of Mathematical Physics* (Nauka, Novosibirsk, 1982) (in Russian); English translation: in American Mathematical Society Translations: Series 2, vol. 130, 1986

124. V. Ledoux, Study of special algorithms for solving Sturm–Liouville and Schrödinger equations. Thesis Universiteit Gent, 2007

125. A.F. Leontiev, *Generalizations of Exponential Series* (Nauka, Moscow, 1981) (in Russian)

126. B. Ya. Levin, Fourier and Laplace type transforms by means of solutions to second-order differential equations. Dokl. Akad. Nauk SSSR **106**(2), 187–190 (1956) (in Russian)

127. B.M. Levitan, *Inverse Sturm–Liouville Problems* (VSP, Zeist, 1987)

128. B.M. Levitan, I.S. Sargsjan, *Sturm–Liouville and Dirac Operators* (Springer, Dordrecht; Kluwer Academic Publishers, 1991)

129. J.A. López-Toledo, H. Oviedo-Galdeano, Reflection and transmission of a Gaussian beam for an inhomogeneous layered medium using SPPS method. J. Electromagnet. Waves Appl. **32**(17), 2210–2227 (2018)

130. B.D. Lowe, M. Pilant, W. Rundell, The recovery of potentials from finite spectral data. SIAM J. Math. Anal. **23**(2), 482–504 (1992)
131. J. Lützen, Sturm and Liouville's work on ordinary linear differential equations. The emergence of Sturm–Liouville theory. Arch. Hist. Exact Sci. **29**(4), 309–376 (1984)
132. V.A. Marchenko, Some questions on one-dimensional linear second order differential operators, Trans. Moscow Math. Soc. **1**, 327–420 (1952)
133. V.A. Marchenko, *Sturm–Liouville Operators and Applications. Revised Edition.* (AMS Chelsea Publishing, Providence, 2011); also: Operator Theory Advances and Applications, vol. 22, Birkhäuser
134. K. Narahara, Soliton decay in composite right- and left-handed transmission lines periodically loaded with Schottky varactors. IEICE Electron. Exp. **11**(23), 1–10 (2014)
135. M. Neher, Enclosing solutions of an inverse Sturm–Liouville problem with finite data. Computing **53**, 379–395 (1994)
136. F.W.J. Olver, D.W. Lozier, R.F. Boisvert, Ch.W. Clark (eds.), *NIST Handbook of Mathematical Functions* (Cambridge University Press, New York, 2010)
137. J.W. Paine, A numerical method for the inverse Sturm–Liouville problem. SIAM J. Sci. Stat.Comput. **5**(1), 149–156 (1984)
138. J.W. Paine, F.R. De Hoog, R.R. Anderssen, On the correction of finite difference eigenvalue approximations for Sturm–Liouville problems. Computing **26**, 123–139 (1981)
139. J. Poschel, E. Trubowitz, *Inverse Spectral Theory* (Academic Press, London, 1987)
140. A. Ya. Povzner, On differential equations of Sturm–Liouville type on a half-line. Mat. Sb. **23**(1), 3–52 (1948) (in Russian)
141. A.P. Prudnikov, Yu. A. Brychkov, O.I. Marichev, *Integrals and Series. vol. 2. Special Functions* (Gordon & Breach Science Publishers, New York, 1986)
142. J.D. Pryce, *Numerical Solution of Sturm–Liouville Problems* (Clarendon Press, Oxford, 1993)
143. V.S. Rabinovich, V. Barrera-Figueroa, L. Olivera Ramírez, On the spectra of one-dimensional Schrödinger operators with singular potentials. Front. Phys. **7**, 57 (2019)
144. V.S. Rabinovich, J. Hernández-Juárez, Method of the spectral parameter power series in problems of underwater acoustics of the stratified ocean. Math. Methods Appl. Sci. **38**(10), 1990–1999 (2015)
145. V.S. Rabinovich, J. Hernández-Juárez, Effective methods of estimates of acoustic fields in the ocean generated by moving sources. Appl. Anal. **95**(1), 124–137 (2016)
146. V.S. Rabinovich, J. Hernández-Juárez, Numerical estimates of acoustic fields in the ocean generated by moving airborne sources. Appl. Anal. **96**(11), 1961–1981 (2017)
147. V.S. Rabinovich, F. Urbano-Altamirano, Application of the SPPS method to the one-dimensional quantum scattering. Commun. Math. Anal. **17**(2), 295–310 (2014)
148. V.S. Rabinovich, F. Urbano-Altamirano, Transition matrices for quantum waveguides with impurities. Math. Methods Appl. Sci. **41**(12), 4659–4675 (2018)
149. M. Rafler, C. Böckmann, Reconstruction method for inverse Sturm–Liouville problems with discontinuous potentials. Inverse Prob. **23**(3), 933–946 (2007)
150. A.G. Ramm, *Inverse Problems: Mathematical and Analytical Techniques with Applications to Engineering* (Springer, Boston, 2005)
151. N. Röhrl, A least-squares functional for solving inverse Sturm–Liouville problems. Inverse Prob. **21**, 2009–2017 (2005)
152. V.G. Romanov, *Inverse Problems of Mathematical Physics.* (VNU Science Press, Utrecht, 1987)
153. W. Rudin, *Real and Complex Analysis*, 3rd edn. (McGraw-Hill, Singapore, 1987)
154. W. Rundell, P.E. Sacks, Reconstruction techniques for classical inverse Sturm–Liouville problems. Math. Comput. **58**, 161–183 (1992)

155. W. Rundell, P.E. Sacks, On the determination of potentials without bound state data. J. Comput. Appl. Math. **55**, 325–347 (1994)
156. P.E. Sacks, An iterative method for the inverse Dirichlet problem. Inverse Prob. **4**, 1055–1069 (1988)
157. J.K. Shaw, *Mathematical Principles of Optical Fiber Communications* (SIAM, Philadelphia, 2004)
158. S.M. Sitnik, *Transmutations and Applications: A Survey* (2010), p. 141. http://arxiv.org/abs/1012.3741
159. S.M. Sitnik, E.L. Shishkina, *Method of Transmutations for Differential Equations with Bessel Operators* (Fizmatlit, Moscow, 2019) (in Russian)
160. S.M. Sitnik, E.L. Shishkina, *Transmutations, Singular and Fractional Differential Equations with Applications to Mathematical Physics* (Elsevier, Amsterdam, 2020)
161. P.K. Suetin, *Classical Orthogonal Polynomials*, 3rd edn. (Fizmatlit, Moscow, 2005) (in Russian)
162. G. Teschl, *Mathematical Methods in Quantum Mechanics with Applications to Schrödinger Operators*. Graduate Studies in Mathematics, vol. 99. (American Mathematical Society, Providence, 2009)
163. E. Titchmarsh, *Eigenfunction Expansions Associated with Second-Order Differential Equations*, 2nd edn. (Clarendon Press, Oxford, 1962)
164. K. Trimeche, *Transmutation Operators and Mean-Periodic Functions Associated with Differential Operators* (Harwood Academic Publishers, London, 1988)
165. C. van der Mee, S. Seatzu, D. Theis, Structured matrix algorithms for inverse scattering on the line. Calcolo **44**, 59–87 (2007)
166. A.O. Vatulyan, *Inverse Problems of Solid Mechanics* (Fizmatlit, Moscow, 2007) (in Russian)
167. A.O. Vatulyan, *Coefficient Inverse Problems of Mechanics* (Fizmatlit, Moscow, 2019) (in Russian).
168. J. Walter, Regular eigenvalue problems with eigenvalue parameter in the boundary condition. Math. Z. **133**, 301–312 (1973)
169. H. Wang, Sh. Xiang, On the convergence rates of Legendre approximation. Math. Comp. **81**, 861–877 (2012)
170. G.N. Watson, *A Treatise on the Theory of Bessel Functions*, 2nd edn. (Cambridge University Press, Cambridge, 1996)
171. J.E. Wilkins, Neumann series of Bessel functions. *Trans. Am. Math. Soc.* **64** , 359–385 (1948)
172. V.A. Yurko, *Introduction to the Theory of Inverse Spectral Problems* (Fizmatlit, Moscow, 2007) (in Russian)
173. V.E. Zakharov, A.B. Shabat, Exact theory of two-dimensional self-focusing and one-dimensional self-modulation of waves in nonlinear media. Sov. Phys. - JETP **34**, 62–69 (1972)
174. A. Zettl, *Sturm–Liouville Theory* (American Mathematical Society, Providence, 2005)
175. F. Zhang, X. Zhou, Capillary surfaces in and around exotic cylinders with application to stability analysis. J. Fluid Mech. **882**, A28 (2020). https://doi.org/10.1017/jfm.2019.824
176. F. Zhang, X. Zhou, General exotic capillary tubes. J. Fluid Mech. **885**, A1 (2020). https://doi.org/10.1017/jfm.2019.982
177. D. Zwillinger, *Handbook of Differential Equations* (Academic Press, San Diego, 1997)

Index

Printed in the United States
By Bookmasters